THE ACCIDENTAL *HOMO SAPIENS*

THE ACCIDENTAL *HOMO SAPIENS*

Genetics, Behavior, and Free Will

IAN TATTERSALL
AND ROB DESALLE

PEGASUS BOOKS
NEW YORK LONDON

THE ACCIDENTAL HOMO SAPIENS

Pegasus Books Ltd.
148 W. 37th Street, 13th Floor
New York, NY 10018

First Pegasus Books edition April 2019

Interior design by Maria Fernandez

Library of Congress Cataloging-in-Publication Data is available.

ISBN: 978-1-64313-026-2

10 9 8 7 6 5 4 3 2 1

Printed in the United States of America
Distributed by W. W. Norton & Company, Inc.
www.pegasusbooks.us

In memory/celebration of Gisela / Tat and Gracie / Bucket,

our parents,

who taught us the value of thinking critically and speaking softly

CONTENTS

FOREWORD

Plac'd on this isthmus of a middle state
A being darkly wise and rudely great . . .
Sole judge of truth, in endless error hurl'd:
The glory, jest, and riddle of the world!

—Alexander Pope,
"Essay on Man, Epistle II"

Human beings will always be an unfathomable mystery. Which is of course ironic, because as members of the world's only introspective species, humans will only ever be mysteries to themselves. There are numerous ways in which members of *Homo sapiens* literally stand out among all the many millions of other living species that populate the globe, most of them relating in some way or other to our unusual upright posture. But none of our physical peculiarities makes us more baffling than a much less material feature: the way in which we process information. Our curious cognitive style places us apart from the rest of nature, not only in our capacity to comprehend and exploit the world in which we live, but in our ability to fabricate and believe reductionist stories about it. In recent years, some of the most seductive of those stories have involved directly imputing our behaviors to our genes, implicitly absolving us of responsibility for some of the more bizarre and irresponsible things we do (and, incidentally, robbing us of credit for the more admirable ones). But while such absolution may come as something of a relief to members of a species that seems inherently conflicted about most of what it does, it comes at the cost of ignoring the scientific complexities involved—and the wonderful fact that, unlike the members of all other living species, human beings actually do have choices.

Hence this book, the fruit of a collaboration that has now extended over a decade, since we co-curated the Spitzer Hall of Human Origins at the American Museum of Natural History. Back then, we discovered a mutual worry about the way in which reductionist stories about who we are and how we got here have tended to grab public attention at the expense of scientific accuracy. Accordingly, this is our attempt to look at the relationship between our genes and our behaviors, and to place the evolution of the human species within the framework of what we know about the evolution of life more generally. For it becomes ever more evident that, no matter how astonishing the result—and *Homo sapiens* certainly is astonishing—the process by which it came about was entirely routine and unremarkable in the grand scheme of life on this planet. Our evolution, as we hope to show, has been far more a story of contingency and chance than of fine-tuning by adaptation; and it is only by sweeping away appealing adaptationist stories and consciously accepting this fact, that we will ever fully make sense of what we have become.

So here is our best effort to understand how we modern humans became the extraordinary and hugely inconsistent creatures that we are, and to explain how and why we will continue to be this way. Along the way we will also touch on what we didn't become, and what we won't become. For, while our species *Homo sapiens* is far from a perfected product, neither does it appear to be a biological work in progress, at least under current demographic circumstances. Whatever we become will be determined not by biological innovation, but by an energetic and unprecedented cultural dynamic. That dynamic is not tightly constrained by the heritage of the past—although, perhaps paradoxically, it appears to be only precariously under our control. Still, as we hope to make clear, if we botch our future it will not have been inevitable.

This book could never have seen the light of day without the vision and enthusiasm of Jessica Case, our editor at Pegasus Books, and of our agent, Don Fehr. Thank you both, and thanks also to Jane Isay and Michelle Press, who have been a constant source of wisdom and encouragement in our efforts to communicate our science. On the production side, we are grateful to Meghan O'Brien for her sensitive copyedit, to Rita Madrigal for her acute proofreading, and to Maria Fernandez for the book's elegant design and for skillfully keeping everything on track. We are deeply indebted to Kayla Younkin, Jay Matternes, Diana Salles, and Ken Mowbray for producing the elegant illustrations; and as always our affection and appreciation go to our wives, Jeanne and Erin, for their support and forbearance.

PROLOGUE

In his novel *The Kites*, set in Europe amid the Nazi barbarities of World War II, the Romanian-French novelist Romain Gary neatly encapsulated the human predicament. "Part of being human," he wrote, "is the inhumanity of it." And for good measure, he added that "as long as we refuse to admit that inhumanity is completely human, we'll just be telling ourselves pious lies."

To anyone who might—against all odds—believe in the perfectibility of humankind, Gary's view of *Homo sapiens* might appear a little uncharitable. But a dispassionate view of history will leave most of us in little doubt about the enduring accuracy of his observation. True, during its tenure *Homo sapiens* has been responsible for a practically inexhaustible list of wholly admirable achievements. But it is nonetheless undeniable that the list of the miseries our species has inflicted on itself, and on the world in general, is impressive enough to suggest that Gary was spot on: that "inhumanity" is just as truthful a descriptor of humankind as "humanity" is. Or, perhaps more accurately, that any adequate characterization of our species requires using both descriptors.

Interestingly, this inconsistency applies at the level of the individual as well as at that of the species as a whole. Knowing that someone is a member of *Homo sapiens* tells us that he or she is an upright biped with clever hands and a big brain. But it tells us nothing about the kind of person he or she happens to be. Indeed, it would be easy enough to find within the species—or even, quite probably, among the human residents of a single city block—individuals who would fit any pair of behavioral antitheses you might care to mention: saintly or selfish; generous or grasping; vindictive or forgiving; wise or gullible; meek or aggressive; energetic or indolent. In fact, any one of us might be any of these things in the very same day, for every human being is a bundle of contradictions.

At the extreme, socially conservative politicians rail publicly against licentious behaviors and reproductive choice, while sending their mistresses out for discreet abortions; pastors preach virtue, then abuse children; depressed comedians kill themselves. What is more, although usually on a smaller scale, contradictory behaviors of this kind are shown by everybody, without exception. Not all of us can be characterized in quite the extravagant dichotomies once used by a book-jacket blurb writer to describe Samuel Johnson's friend and biographer James Boswell as "loving husband and father, dissipated whore-chaser; conscientious lawyer, drunken buffoon; writer of tedious doggerel, and author of one of the finest biographies in the English language." But in every one of us these or similar words will strike an echo of some kind, however loud or faint it may be.

This paradoxical nature of ours is why what defines us as human, the "human condition" that philosophers and others have so long striven to characterize, is maddeningly elusive; and it is supremely ironic that the only species that—so far as we know—agonizes about its condition turns out to be the very species in which that quality is hardest to pin down. Of course, it is entirely true that we share a very great deal of our cognitive as well as physical and genomic makeup not only with our closest relatives in the living world, the great apes, but with a vast, radiating array of forms far beyond them. But there is nonetheless something emergently *different* about the way in which we process information in our minds. And that disparity has fundamentally changed the rules by which we interact with the rest of the living world—and with each other, for that matter. The difference, and how we acquired it, is what this book is about.

As evolutionary biologists, we are acutely aware that human beings are intimately nested into our planet's biosphere: that we are related by ancestry to all the living things we see and don't see around us. We *are* organisms, after all. But it is also evident that we humans do business in the world in a qualitatively unique way, and that there is a deep—even if narrow—behavioral gulf between us and *all* other living things, even including those very close relatives, the apes. What makes us so different is that we have become *symbolic*. This is a shorthand way of saying that, to the best of our knowledge unlike any other creatures, we dissect the experience of our internal and external worlds into a vocabulary of mental symbols. We can then mentally combine and recombine those symbols, according to rules, to produce new visions of those worlds. In the pages to come, we outline the evolutionary and genomic mechanisms by which this difference came about, and we briefly recount the long and absorbing history of how we acquired that uniqueness. And we try to understand—as much for our own benefit as for yours—why members of our species so often behave in bizarre ways, while always remaining recognizably human—at least in Romain Gary's sense of the word.

We are, of course, very far from the first to attempt the hubristic enterprise of explaining just what it is that makes humans different. Philosophers, dramatists, and poets have striven for millennia to encapsulate, if not to explain, our bizarre experience of ourselves. And with a gradual shift in literary tastes, over the last couple of centuries it is the novelists such as Gary—and in recent times moviemakers—who have become incomparably the most successful characterizers of the unique human experience, albeit necessarily capturing it fragment by tiny fragment.

Novelists can, of course, allow their imaginations to roam freely. Scientists, who came relatively late to the game, labor under substantially tighter constraints. And they have generally found that making solid generalizations about the ways in which *Homo sapiens* individually and collectively behaves is an intractable task at best. On the clinical front, for example, from which *Homo sapiens* is usually seen not as dynamically evolving but as a biological fait accompli, psychologists and psychiatrists have tried to chart, explain, and organize deviations from expected behavioral norms. The enormous debate this effort has unleashed may reflect to some extent the awkward fact that humans are deeply cultural creatures, and that those norms are at least as much learned cultural artifacts as they are inherited biological features.

In the domains of anthropology and behavioral biology, the scientists who call themselves evolutionary psychologists have done a great deal to document cross-cultural and other regularities in human behavior, in pursuit of the notion that our modern behaviors are mediated by ancient influences. But they have greatly handicapped themselves by examining specific behaviors such as altruism, or infidelity, or homosexuality, as individuated and inherited expressions that can be explained in isolation. That disadvantage comes about because, although cherry-picking discrete items in this way makes analysis easier, every behavior actually exists on a spectrum. Such spectra are defined by what statisticians call the "normal distribution," more popularly known as the "bell curve." In any normal distribution, a few individuals mark the extremes of the spectrum (running from greed to generosity, say), while the bulk of the population falls somewhere in the middle (most of us being for the greater part reasonably open-handed, even if not always).

The omnipresence of the bell curve affects the way in which we need to look at the expression of human behaviors. Take altruism, probably the evolutionary psychologists' favorite behavior. This, the propensity of certain individuals to sacrifice themselves for the benefit of others, may look at first glance like an extraordinary manifestation that begs for a special explanation. But when you realize that altruism is symmetrically matched by extreme selfishness at the other end of the distribution, and that most of us lie somewhere in between, something else emerges. Namely, that what is most worthy of investigation is not the individual extremes, but the nature and existence of the curve itself. And that curve is a property of the species, not of the individual—which is one major reason why their common membership in *Homo sapiens* predicts notably little about its individual members.

Given all this, it is hardly surprising that the human condition has been so elusive. One approach to solving the conundrum of what makes our behavioral repertoire unique, and one that evolutionary psychologists have particularly favored, is the making of laundry lists of "human universals." These are things that all humans do (or could do), but that nothing else does. On the face of it, the human universal perspective certainly appears to provide a reasonable starting point for understanding whatever it is that makes us unique. After all, isn't what makes us different merely the sum of the things that only we do? But look a bit closer, and you see that this approach doesn't get us very far in any meaningful descriptive sense. For example, our explicit awareness of our impending mortality is almost certainly unique to human beings; but even though this knowledge may do a lot to explain our general existential angst and its many derivatives (just as the knowledge of their immortality oppressed Gulliver's Struldbrugs), it is hardly a satisfactory predictor of our individual dispositions. And no matter how long your list gets, it will never be comprehensive (there will always be something you forgot, rather like thinking of the largest number you can imagine, and then adding one), while at the same time it will always be riddled with exceptions. That damned curve again.

Of course, there can be little doubt that—to a certain modifiable extent—every individual is the sum of his or her genetic predispositions; and most of us, at least, are born more or less the kind of individual we are likely to remain, unless we make a conscious and often difficult effort to behave contrary to our basic instincts. But, while it may be true that close observers can bracket the behavior patterns of individuals with some accuracy, when we seek a defining behavioral condition for our species as a whole we are nonplussed. Statistics and the curve aside, the main reason for this difficulty is that, while the most remarkable and consequential of our unique and universal human capacities is our common ability to reimagine the world, we each do so in our own particular way. So much so that, in our intensely social species, culture and its rules play critically important roles in keeping individual worldviews reasonably compatible among the members of each society.

Still, that cultural containment comes at a price. We all come into the world with the potential to absorb any language or set of cultural norms that *Homo sapiens* has to offer; yet, by an early age, we may have absorbed an unshakable perspective on the world that is completely incompatible with that of members of other societies (or even, occasionally, of our own). To put it another way, our symbolic capacity makes it possible for individual human beings to live in subjective worlds that are far more distinct from those of other humans than are the more immediate and practical worlds inhabited by, say, members of different ape species. And in an age of globalization, the full consequences of this particularity of ours are becoming ever more apparent.

Given all these complications, it is inevitable that the human condition should remain elusive; and if we were pressed to think of the single human universal that best embraces our condition, we would reluctantly have to choose "cognitive dissonance," although this may of course be a result of the current state of our national politics. Yet, there is equally no doubt that we human beings are the product of a long and complex evolutionary process that has in some way tailored us to the world we inhabit; and, at least in certain respects, we *are* captive to our unusual human biology. This paradox inevitably haunts the center of this book, in which our main aim is to examine the interface between our biological inheritance and our behaviors, and to discover just how we contrived to become the wonderful, flawed, infuriating (choose your own adjectives) creatures that we are today.

ONE

GENES, EVOLUTION, AND THE BELL CURVE

Human beings are intimately nested into the natural world. Because of this they share many features with other multicelled organisms, even while differing from all of them to widely varying degrees and at numerous levels of complexity. Because the great tree of life to which we belong is literally a genealogy, these similarities and differences are hierarchically arranged. This organization makes it possible to gather information at many different levels in pursuit of understanding why both the similarities and the differences exist. One way to start is by looking at very highly integrated—and even emergent—levels of organization, such as culture and group behavior. But we can, if we wish, move one level lower and look at whole individuals. Or we can dig a little deeper and look at the organs of our bodies. Or we can proceed to an even finer level and focus on the tissues that make up those organs. But that's not the end of it. At a lower level yet, we can examine the cells that make up the various tissues. Deeper still, we can look at the kinds of structures of which cells are composed, such as nuclei and membranes; or we can crank up the magnification yet more and look at the molecules from which our cells are built. And it is at this most intimate of levels that we best see the basic repetitive patterns that bind together every living organism on Earth.

Even in this elemental perspective, it becomes quickly apparent that we are not the only complex beings on Earth. Indeed, even very simple-looking single-celled bacteria exist at a level of complexity that is nothing short of exquisite. Yet, for all the many ways in which this complexity is expressed across the many branches of the natural world, there is a single unifying theme that underlies everything we see. Namely, that it is the job of all organisms, whatever their means of doing business, to make more of themselves. And this is a task that obviously demands a mechanism for faithfully replicating both organisms, and their traits, from one generation to the next.

On our planet, this reproductive mission is accomplished by using the most basic units of our being: molecules, more particularly, protein molecules. Every part of us and of every other living organism—our bodies, behaviors, and, in the human case, even our beliefs—has a molecular basis. And, as a result, there is a very real continuity between the simplest and the most complex aspects of what we are. As the great molecular biologist and polymath Francis Crick once said, very explicitly, "A person's mental activities are entirely due to the behavior of nerve cells, glial cells, and the atoms, ions, and molecules that make them up and influence them." In other words, even our thoughts and emotions are in essence nothing more than the sum of a host of molecular and chemical reactions. What is more, if whatever generates our mental capacities ultimately resides at a molecular level, then so must the underpinnings of the emergent properties of those mental capacities, such as culture and group behaviors. This is a truly sobering thought; and it is why we are beginning this book by discussing the fundamentals of the tiny structures that are so exquisitely combined to make us what we are.

All living organisms on this planet have arisen from a single primordial common ancestor that existed perhaps as much as 3.5 to 4 billion years ago. All subsequent life sprang from this ancestor, branching and diverging to ultimately produce the many millions of different kinds of organisms that exist today. We know this not only through the meticulous analysis of anatomical structures across this vast array of creatures, but because we have lately accumulated a remarkable knowledge of the genomes of organisms—not only from all existing branches of the tree of life, but even from some extinct forms. The structural similarities among those genomes clearly confirm a pattern of continuity and relatedness among all living things on Earth.

We see this perhaps most clearly in two things. First, the proteins that organisms are composed of are not organized randomly. Rather, they are fine-tuned molecular structures, built on a common pattern, that perform with exquisite precision to assure the survival of each organism, whatever its external appearance. Second, the blueprints for these proteins are all embedded in the amazing linear DNA molecules that make up the genome. And since genomes are an important part of understanding many of the nuances of the human condition, we need to spend a little time exploring their basics.

PLAYING WITH LEGOS

Nature seems to have settled on a linear theme (long strings of small, basic components) at both the molecular and the cellular levels. But molecular variety is not confined to linear patterns. Some molecular structures are twisted so that they assume fantastic shapes and substantial three-dimensional complexity; and indeed, their three-dimensional shapes are essential to the ways in which they function.

Researchers have found over ten million different proteins in the cells of the many organisms that have been studied so far. This is a substantially large number, and an even greater number of cellular functions has been identified. To coordinate all this complexity of function, every cell needs a blueprint for each protein it manufactures. What is more, for the cell machinery to run smoothly from one generation to the next, that blueprint not only needs to be inherited in a smoothly integrated fashion, but it also needs to be able to reliably replicate itself. Yet, if evolutionary change is to take place—and evolution is the only known mechanism that actually predicts the diversity of life we see around us—there must also be room for changes to occur between generations. Nature has found an incredibly efficient solution to these diverse requirements by deputing the job of maintaining and transmitting those blueprints to a very specific kind of molecule known as deoxyribonucleic acid, or DNA. The structure of DNA is not only well suited for holding information, but also for self-replication; and indeed, the DNA molecule is a thing of amazing beauty for its utility as much as for its structure. Playing with LEGOs can help us understand the amazing capacity of DNA to hold the diversity of information necessary for all life on our planet.

A basic LEGO set has six building blocks, of six different colors, with eight pegs on each one. Any parent knows (in addition to the excruciating pain caused by stepping on a LEGO in the middle of the night) just how complex the structures that a four-year-old kid can create may be with this tiny number of different blocks. In fact, a Danish math professor named Søren Eilers became famous for computing the exact number of permutations that these six differently colored blocks can produce: an astonishing 658,874,710,800. These nearly 660 billion permutations include a lot of combinations with the same shape, but with different colors distributed among them. (The original calculation was exactly 915,103,765, but Eilers did not compute the possibilities considering the six different colors. With colors added, there are 720 times more combinations.) So, if we allowed two four-year-olds to independently make structures with this basic six-piece LEGO kit, we would undoubtedly see pretty divergent end products. Each kid would use different colors and units, and would connect the blocks differently, although if we were to limit them by specifying that they could make only lines of blocks, there would be a mere 1 in 720 chance that the kids would make the same arrangement of colors.

So how do these rules apply to DNA and proteins? The long DNA molecule has four basic building blocks (called nucleotides), and proteins have twenty or so basic building blocks (called amino acids). The number of ways in which four different things can be arranged in a line is a mere twenty-four; but the number of ways in which twenty different things can be arranged in a line is 2,432,902,008,176,640,000. Even with just four basic units, there is an unimaginable number of different ways in which we can arrange things linearly using longer strings. For instance, if we consider a ten-unit-long sequence of DNA, the number of possible arrangements is only 410; but if the string is allowed to be 100 bases long, there are 1,099,511,627,776 ways to arrange the four things. For a protein with twenty amino acids, there are 10^{131} (that's a 1 with 131 zeros after it) ways to arrange 100 amino acids in a linear pattern.

Most proteins are 300 to 500 amino acids in length, so the number of possible combinations of amino acids in a typical protein is 10^{390}. And for a DNA sequence of, say, 999 bases, the possible combinations are 10^{601}. (DNA that codes for proteins comes in multiples of three, and there are nearly three times the number of bases in a gene coding for a protein as there are amino acids coded for by that gene.) Clearly, nature has stumbled upon a unique and virtually inexhaustible way of storing information using linear DNA and protein sequences.

Deoxyribonucleic acid is made of smaller molecules and atoms that come together in an elegant confluence of structural rigidity, information, and functionality. The DNA molecule's structural rigidity arises from the way the atoms and the smaller molecules are arranged and bound to each other to comprise it. We knew long before the discovery of the DNA molecule's structure in 1953 that it is made up of carbon, hydrogen, nitrogen, phosphate, and oxygen atoms: five of the smaller atoms in the periodic table. And now we know that in DNA these atoms come together to form four kinds of larger building blocks. These four blocks share a basic core structure, but they are differentiated from one another by side chains of molecules known as "bases." The four bases are guanine, adenine, thymine, and cytosine or, in DNA lingo, G, A, T, and C.

The DNA molecule itself takes the form of a double helix, in which two antiparallel (meaning parallel but running in different directions) linear molecules contact and intertwine with one another. It turns out that the double helix is very stable, staying intact even in the face of high temperatures and noxious chemical environments. This is because the bases in one of the linear molecules bind solidly to the bases in the other. But the bonds aren't random. G on one strand always binds to a C on the other, and a T on one strand always binds to an A on the other. And what this means is that if you have one of the linear strands, you know the structure of the other. You can thus re-create it by simply letting the strand you have bind to the right bases, and then connecting the new complementary bases to build the new DNA strand. This connection process is known by biologists as DNA replication, and by chemists as polymerization.

The actual task of replicating the DNA is accomplished by several molecules. A molecule, aptly named a helicase, unwinds the tangled double helix. Another, called DNA polymerase, takes each of the two strands of the double helix separately and, by using basic chemical reactions, reconstructs their complementary strands. This happens every time a cell divides, and because DNA resides in both the ova of the mother and the sperm of the father, DNA is the perfect medium for transferring genetic information from one generation to the next. Still, if the copying mechanism were entirely reliable there would be no way of introducing the hereditary changes that we know characterized the evolution of life on Earth. Fortunately for the evolutionary process, that mechanism is very slightly imperfect; despite the overall high efficiency of the replication machine, the

DNA polymerase that does the heavy lifting in DNA copying does make rare mistakes. Those mistakes, known as mutations, are then passed on to the next generation when a sperm and the ovum fuse.

Once passed to the next generation, there are three possible effects of the mutations. In the first and simplest instance, mutations will have no effect at all, in which case they are neutral. The second category of mistakes includes those that are detrimental to the unfortunate individuals who have inherited them. In the worst of these cases such mutations are lethal, or cause severe genetic disorders, and will be rapidly weeded out of the population. The third—and most intriguing—possibility is that a mutation will give its possessors an advantage of some kind. Under the right circumstances, such advantageous mistakes will be favored, yielding the potential for evolutionary change with the passage of time.

DOGMA, CODES, AND RULES

The basic information for making proteins is contained in DNA. But something else is also involved in the making of proteins, because what the DNA blueprints actually do is to specify an intermediate molecule that does the real hard work: ribonucleic acid, or RNA. The type of RNA involved is called messenger RNA (mRNA) because it relays the message of the DNA to the cell for translation into protein. This molecule has close structural similarities to DNA, and it uses some of the same basic units. The RNA and DNA molecules work in tandem, giving rise to the "central dogma of molecular biology," namely that DNA is *transcribed* to RNA, which is *translated* to PROTEIN. All living things on this planet abide by this dogma (except for some viruses, but that's another story). The hierarchy of DNA, RNA, and proteins means that there are extra moving parts in the cellular system, but it still works very efficiently.

One of the implications of the central dogma is that, for information to flow from DNA to protein via the RNA, the DNA must code for the proteins that make up our bodies. But there is a coding problem, because while there are twenty amino acids to specify, there are only four basic DNA building blocks: G, A, T, and C. A one-to-one code could accommodate only four amino acids, while even using pairs of nucleic acids would leave us short. To illustrate this, let's go back to our LEGOs and start with several blocks that have four different colors: red, blue, green, and yellow. Try to arrange two blocks in every possible color combination: start with red, red, then maybe red, green, and then red, blue, and so on. If you do this, you will quickly discover that there are only sixteen ways to arrange two blocks with the four different colors. And if the four LEGOs with different colors represent G, A, T, and C, then it is obvious that the sixteen different pairs of colors (known as "codons") still leaves us four amino acids short. Clearly, we need to go to triplet codons that consist of three bases. But if we do this, we have the opposite problem, because there are sixty-four ways to arrange four colors in groups of three, leaving us with a surplus of forty-four codons.

Nature could have solved this problem in any number of different ways, but it did so by taking advantage of redundancy: that is, by having multiple codons for the same amino acid. Our favorite example of this is the four codons that code for the amino acid proline (also known as P or pro). There are four codons in DNA that can code for proline: CCC, CCT, CCG, and CCA. Four other amino acids are also specified by four different codons, nine are specified by two, one by three, two by only one, and three are coded for by six different codons. If you have been counting, this accounts for only sixty-one of the sixty-four codons that a three-letter code can accommodate; and in a truly economical stroke, the genetic code uses the missing three codons to tell the protein-making apparatus to stop. This additional flourish is beginning to make the genetic code look like a written sentence, with three-letter words that translate into amino acids, and a period at the end. But we still need to capitalize the beginning of the sentence, and the code does this by designating the codon for the amino acid methionine (ATG) for this role. So, let's imagine an imaginary DNA sequence in which one of the two strands looks like this:

atcagctgacatcgagcctgccatgccaccgcccctcctccacccctccataagacacg

On the face of it, this looks more like a garbled string of letters than a sentence, because we don't see certain linguistic characteristics that we expect of a sentence printed on a page. But the molecular rules we have described above actually do give it a sentence-like structure. First, our rules say that the capitalization of the sentence should occur at the codon ATG. Scanning the sequence, we find our capitalization at the first and only ATG:

atcagctgacatcgagcctgcc**atg**ccaccgcccctcctccacccctccataagacacg

Finding this ATG does two things to help us read our protein sequence. First, it shows us where the first word in the sentence is. And second, since we know that all the words in the sentence are three letters long, it also allows us to tease them apart. If we disregard the letters to the left of the ATG, we can now write the sentence as:

atg cca ccg ccc cct cct cca ccc cct cca taa gac acg

So now we know where our sentence starts, and where our words are. Our next step is to use the genetic code to translate to a protein sequence. What we show below is the DNA sequence with the triplet codes translated to amino acids. Remember also that CCC, CCG, CCT, and CCA all code for P, or proline.

M	P	P	P	P	P	P	P	P	P	.
atg	cca	ccg	ccc	cct	cct	cca	ccc	cct	cca	taa

Through this step we finally discover where the sentence ends, since TAA codes for a stop. Even though they code for amino acids, the last triplets are irrelevant to the protein sequence, and as a result we find that what we have here are the instructions for a strange protein that contains almost all prolines.

One important outcome of having redundancy among codons is that, while the DNA sequences might spontaneously change over time through mutation, the proteins they specify will not necessarily do so. Compare these two sequences:

Sequence 1	cca ccg ccc cct cct cca ccc cct cca
Sequence 2	ccc ccc ccc ccc ccc ccc ccc ccc ccc

Both sequences code for a protein with nine Ps, but there are seven changes between them. Changes of this kind are called "silent" because they have no effect on the protein produced. Researchers can use DNA sequence patterns of this kind to determine if a protein has been influenced by natural selection, and this will become important when we discuss the evolution of some of our human traits. Any protein sequence that remains the same will normally be under strong pressure from natural selection to stay the same; but it may nonetheless incur a lot of those silent DNA sequence changes. On the other hand, any DNA sequence that codes for a protein that is not influenced by natural selection should have about the same number of silent changes as it does DNA changes that materially alter the amino acid sequence ("replacement" changes). The ratio of replacement to silent changes then gives us a metric for the presence and intensity of selection acting on a protein. If natural selection is neutral for a given protein, then the ratio should be 1.0, the number of replacements being equal to that of silents. But if natural selection is actively purging replacements, then the ratio should be less than 1.0. And if the ratio is greater than 1.0, the indication is that natural selection is favoring change.

All this might seem like a lot of detail for a book that aims to understand the big picture of what it means to be human. But because DNA is so important to what you are—and to what your ancestors were—it is really helpful to understand the basics.

A VERY SHORT COURSE IN GENETICS

As eukaryotes (complex organisms that have their genomes encased in a membrane within the cell), human beings reproduce by combining genetic material from both male and female parents. In other words, humans have sex. For better or worse, most organisms on this planet, including the single-celled bacteria that make up 90 percent of all the species in the biota, do not have sex; and those that don't contain in general only a single copy of each DNA-containing chromosome in each cell. They reproduce by replicating their genomes and then simply splitting into two cells; and it is during the replication process that the mutations occur that supply the variation necessary for evolution. In contrast, eukaryotes start out with two copies of each chromosome, reduce that number to a single copy when they make either sperm or eggs, and return to two copies when a sperm and an egg unite. The mutations necessary for evolution occur during the process of "meiosis," when the sperm and egg are formed.

Sex provides an additional way to generate inherited variations among individuals in a population. Variation of this kind is highly desirable, because without it all members of the population would be equally subject to external influences, rendering all of them equally vulnerable to changes in the environment. For example, a familiar supermarket staple, the Cavendish banana, faces imminent extinction (like its predecessor the Gros Michel before it), because cultivated bananas reproduce clonally. This means that all Cavendish plants are genetically identical, and thus equally susceptible to the deadly Tropical Race 4 disease that is currently ravaging plantations in Africa and Asia. If some individual Cavendish plants were less susceptible than others, then these more resistant forms would rapidly come to dominate under this severe selection, and thus avert extinction. In addition, the very process of genetic mixing via sex opens a route to new genetic combinations. Because of the way in which the DNA is packaged in chromosomes, the two parental sets of genetic material have the potential to recombine as they intermingle. How this happens is key to understanding replication, variation, and, eventually, evolution.

One of the most impressive intellectual advances in understanding life on this planet was the painstaking clarification of the mechanisms of heredity by such nineteenth and early twentieth century luminaries as Gregor Mendel, Thomas Hunt Morgan, and H. J. Muller. Right up to the middle of the last century, the "genes" themselves could not be directly visualized. Instead, their existence had to be inferred from observations of the physical traits that they influenced. The genius of the early pioneers of genetics lay in choosing what are now known as "Mendelian" traits: simply inherited features in which, for practical purposes, one gene corresponded to one characteristic. This allowed them to study the independent behaviors of single genes, or the recombining behaviors of a handful of genes, by the observation of phenotypes.

Early in the study of genetics, the discrete characteristics that were studied followed the rules of inheritance closely. Traits such as green versus yellow seeds in peas, and red versus white eyes in fruit flies, provided good examples of simple Mendelian inheritance that involved single genes. This was lucky, because in most cases inheritance is not that simple. As we'll see in the next chapter, most traits of interest to evolutionary biologists are controlled through the interaction of numerous genes, and most genes are involved in influencing more than one feature. An oversimplified example, however, will help demonstrate how such more complex traits behave.

First, let's look at a simple Mendelian trait that is under the control of a single gene. Actually, there may be no such thing in the real world, because while some traits may appear to be under the control of single genes, every trait actually involves the interaction of more than one gene product. In other words, while the effect of a single gene might be huge, the expression of the phenotype (the appearance of the individual) nonetheless also involves RNA polymerases, ribosomal proteins, and other products that are affected by other genes. However, in "discrete" traits the influence of small-effect entities is so minuscule that they can effectively be ignored, and for the purposes of establishing genetic principles the logical—and acceptable—norm is to go with the assumption that certain traits are effectively controlled by a single gene.

Let's assume, first, that the genes we want to examine reside in organisms that reproduce randomly and, second, that in the population concerned there are two different versions (alleles) of the gene. Let's call the alleles r and R). The alleles interact with each other additively (i.e., in genetics-speak there is no "dominance") and so the two "homozygotes" RR and rr have extreme phenotypes, while the "heterozygote" Rr expresses an intermediate phenotype that combines the two. Imagine a system in which RR homozygotes are red, the rr individuals are white, and the Rr heterozygotes are pink. If we were to cross an RR individual with an rr individual, we would get all Rr, demonstrating the intermediate phenotype of this combination. Using the rules of simple Mendelian genetics, in the second generation, where we cross Rr with Rr, we would obtain an equal number of each extreme genotype, and hence an equal number of red and white individuals. But there would be twice as many of the heterozygotic pink phenotype.

Now, let's consider that color is instead controlled by two genes, the simplest case of a polygenic trait. In this case, both genes have two alleles (again R and r for one gene, but now we have D and d for the second gene) and let's also assume that these alleles interact in an additive fashion. The numbers of R and D alleles determine how deep the red of the phenotype will be, such that the RRDD genotype will give the deepest red phenotype. RRDd, which has three different alleles, would have a lighter red phenotype, and so on. Nine different genotypes can be generated for two genes with two alleles each, in this case RRDD, RRDd, RrDD, RrDd, rrDD, RRdd, Rrdd, rrDd, and rrdd. Using our additive rules for the phenotypes, they would look something like this:

Deep red	red	pink	light pink	white
RRDD	RRDd, RrDD	RrDd, rrDD, RRdd	Rrdd, rrDd	rrdd

It should be noted here that not all allelic interactions are additive, and not all genes have just two alleles. But we can nonetheless use these hypothetical genes not only to demonstrate the way that traits can blend, but more importantly, to show how polygenic traits will be distributed in populations.

Finally, if we consider a trait with many (N) genes involved in its expression, we would get the following:

Deep red	range	white
RRDDCCEE . . . NN	. . . . . .	rrddccee . . . nn

The dots in the middle "range" category represent all of the combinations of the alleles that have at least one heterozygote (like RrDDCCEE . . . NN or RrDdCcEe . . . Nn). The more capitalized alleles in a genotype, the redder the trait will be. The more lower-case alleles in the genotype the whiter the trait will be. Genotypes with fifty percent upper case alleles and fifty percent lower case alleles will be pure pink, with all other heterozygous genotypes "ranging" from red to white. This would provide the distribution of phenotypes in the graph below. We will return to the nuances of this distribution later in this chapter.

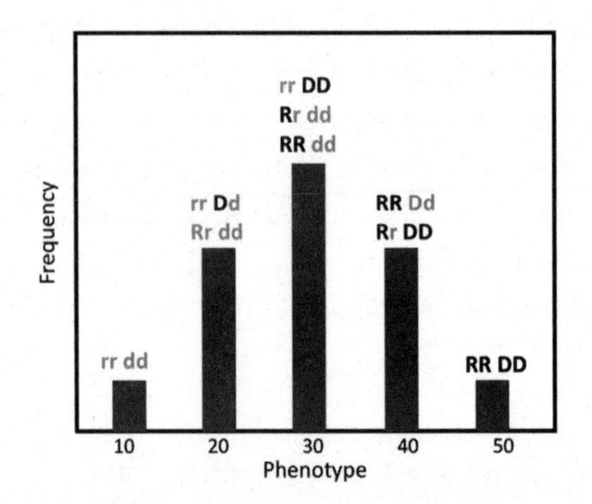

Two-locus, two-allele system where alleles are additive. Black alleles have a positive effect while gray alleles have a negative one, so that the "rr dd" genotype has the lowest fitness, while the "RR DD" genotype has the highest.

You might be wondering how all these different combinations of alleles might be generated. The answer is that the initial variability is generated by mutation, but the way in which the alleles are actually combined is subject to the luck of the draw. In a randomly mating population, for a single gene with two alleles r and R, the outcome will be dictated by probability. Thus, if an RR mother mated with an rr father, the mother would make only R eggs, the father would make only r sperm, and the resulting offspring would necessarily be Rr. However, if an Rr mother were to mate with an Rr father, the offspring would be RR, Rr, and rr in roughly the ratio of 1:2:1, as a result of simple probability. For similar reasons, for two genes (say, the same R and D genes above) with two alleles each, there are nine ways in which the alleles may combine (see the figure above).

The probabilities of getting each of these nine different combinations are well established; and in a randomly mating population, absent any other forces (like natural selection), these probabilities will always express themselves. Geneticists have taken advantage of the regularity of these probabilities to understand how genes are distributed in the genome. Remember that our genome is subdivided into chromosomes, and specific genes lie in specific orders along those chromosomes. The distribution of the randomly mating combinations is used as a baseline to assess if other forces are acting on the genes involved, and it turns out that for genes that are on the same chromosomes and close to each other, these probabilities will be distorted proportionately to their closeness. By looking at recombination rates in offspring from specific kinds of matings, and at how they depart from the baseline of randomness, researchers were able to determine which genes lay close to each other on the same chromosome, and in what order they lay.

Once the basic rules of genetics had been established—namely, that parental traits do not "blend" but instead are determined by genes that are passed along as discrete entities from one generation to the next—the next breakthrough was, ironically, to steer away from simple inheritance, and to recognize instead that most of the more interesting traits in nature are not inherited through the action of single genes, but rather through the joint action of many genes working together. Eventually, this discovery led the evolutionary theoretician Ernst Mayr to dismiss the groundbreaking work of the early twentieth century geneticists as "beanbag genetics." But while Mayr was correct that most characters are not specified in this way, his airy tone was hardly justified. That's because by keeping each gene (bean) in the bag separate from all the others, scientists could easily study their behaviors and begin to model them mathematically. It is hard to see how the scientific study of such an elusive and extremely complex mechanism as inheritance could have begun in any other way.

Fortunately, the development of a field called quantitative genetics shifted our understanding of how physical traits are inherited on to firmer ground. Quantitative geneticists saw traits in organisms as products of complex interactions. Genes can have either additive or dominant effects on traits. But such interactions among the genes themselves are not the only ones that quantitative genetics considers. The most basic quantitative genetics expression is $P = G + E$, where P is the physical phenotype, G is the genetic contribution, and E is the crucial environmental component of the trait. Because the last two of these vary, the phenotype (P) emerges not as a single physical state, but rather as a range of states. By recognizing this, quantitative genetics is able to tackle traits that vary continuously. And, more recently, modern genomics and genome sequencing have begun pointing to the additional sources of complexity we discuss at length in the next chapter.

HEREDITY AND THE ENVIRONMENT

If any one of us had been kept in a black box early in our lives, we would have become something very different from what we are today. This is because early experience is crucial to the outcome of human development, and indeed to the development of any complex being. One of the authors was on a public panel at which the possibility of cloning a Neanderthal was raised. He was appalled by the idea, and the question that most immediately came to mind was, "How would we know how to raise a Neanderthal to be a Neanderthal?" Those extinct ancestors were, after all, highly sentient beings; and, like us, a very large part of them was undoubtedly what they had learned to be.

The psychologist B. F. Skinner made a reputation for himself while still a graduate student at Harvard during the late 1920s, and eventually became one of the most influential behavioral scientists of his generation. Throughout his career, he designed ingenious experiments in which he could control as much of the external environment as possible so that he could test hypotheses about his subjects' behaviors. A "Skinner box," a kind of enclosed chamber for mice or rats with levers inside, can still be purchased from several scientific supply companies—although, since Skinner didn't want his name to appear on such boxes, they are advertised as "operant conditioning boxes." The "operant" part is the unfortunate rodent that is placed inside the box, and the "conditioning" is some outside influence imposed on the hapless subject.

Contrary to legend, Skinner's daughter Deborah did not suffer from her stay in a chamber that he also developed for human babies in the general population. It was often rumored that Deborah grew up to be psychologically damaged, eventually suing him and committing suicide as a result of her stay in this hellish contraption. The reality is different: Deborah grew up quite normally and continued a loving relationship with her father until his death in 1990. The so-called box had simply been a crib known as a "baby tender," which had Plexiglas around it to promote safety and temperature control. Apparently, Skinner's wife had believed that it was best for children to be at a constant ambient temperature, and to be enclosed to protect them from falling. That makes sense to anyone who has recently been around a two-year-old; but such was the scary potential of the Skinner box, that rumors ran out of control.

Assessing the environmental component of any trait requires that you compare its phenotypic trajectory in different environments. The one trick to this procedure is that the genes, or more precisely the genomes, of the two organisms need to be as closely related as possible, preferably identical. With certain plants this can be readily accomplished, because they can be cloned easily. But with animals, making exact genetic animal clones is not easy; indeed, until recently it couldn't be done at all. And while it is possible now, the procedure remains very expensive and is rarely used in experimental biology. What's more, while it is in principle possible to produce a human clone, the ethical ramifications are insurmountable. Still, this has not stopped researchers from studying the impact of environment on human traits because natural human clones do exist, in the form of identical or "monozygotic" (MZ) twins. And, very conveniently, nature also provides a control for identical twin studies, in the form of fraternal or "dizygotic" (DZ) twins. Twin studies have been a mainstay in human genetics for many years.

Scott and Mark Kelly are identical twins. They are one pair of identical twins out of about ten million pairs worldwide. But they are hardly run of the mill. In 1996 both were picked to be astronauts and pilots in the space shuttle program, and they are still the only identical twins to have both traveled in outer space. In March of 2015, Scott was sent into space for a one-year stay on the International Space Station, while Mark remained earthbound for the year. The mission provided a unique opportunity to study the impact of a specific environment—in this case zero gravity—on physiological, mental, genomic, and other factors. One of the more interesting results of the study was that, as soon as Scott entered outer space, gene expression in his body became radically different from the gene expression in his earthbound twin. Changes in the expression of genes for bone formation were among the most notable. In addition, Scott emerged from the flight two inches taller than when he started (Mark did not grow taller during the time span). In this unique experiment, Scott's physiology and genome were both clearly responding to the shock of the new environment.

Classic twin studies are done using a different strategy. MZ twins arise from a single fertilization event; but the two-cell embryo or the morula (what the embryo becomes at a later stage of growth) splits into two new developing embryos, both independently viable. DZ twins, in contrast, are simply the result of two distinct eggs being fertilized by two different sperm in a double fertilization event. MZ twins share all of their genes because they develop from the same fertilization event, while DZ twins share half of their genes just like any biological siblings. If the twins are reared together, the environmental component of their experience should theoretically be a neutral factor in their development, although there are actually two kinds of environmental contribution that can impact the phenotype. The first is the shared environmental component, and the second is the unique environmental component. Contributions from the shared environment include shared prenatal treatment, socioeconomic level, educational factors and the like, while unique environmental impacts come from random events like a broken bone or a prolonged sickness due to infection.

If a trait has a large genetic determinant, MZ twins will show greater resemblance than DZ twins. The degree of concordance is measured using correlation, which is usually assessed on a 0 to 1 scale with 1 being complete correlation, and 0 being a complete lack of correlation. By measuring the difference in the correlations between MZ and DZ twins for a given trait, one can quantify the genetic contribution to that trait. If a trait is entirely controlled by genes, then the correlation for MZ twins should be 1.0 (MZ twins share their entire genomes) and the correlation for DZ twins should be 0.5 (DZ twins share half of their genes). If the shared environment is all that is involved, then the correlations for both MZ and DZ twins should be 1.0. On the other hand, if the unique environmental impact is all that is involved, then the trait measures should be random and have 0.0 correlation for both MZ and DZ twins.

As a real example, consider clinical depression in which correlation measures are 0.4 and 0.2 for MZ and DZ twins, respectively. The difference is 0.2, but this needs to be multiplied by 2 (MZ twins are twice as related to each other as DZ twins are) to quantify the genetic contribution. In this case the corrected difference would be 0.4, which is also the degree to which genes control the trait and is thus what geneticists call the heritability, or h2, of the trait. In 2015, Danielle Posthuma and colleagues conducted a meta-analysis of nearly 3,000 publications from the last sixty years that used classical twin study design. This meta-analysis included 14,558,903 twin pairs (MZ and DZ combined) and examined nearly 18,000 traits. The resulting estimates of heritability revealed a significant genetic contribution for a majority of the traits. The researchers grouped studies and h2 according to what they call "functional domains" such as the psychiatric, metabolic, cardiovascular, or cognitive (there were twenty-eight functional domains in all for those eighteen thousand traits, with the psychiatric domain being the most popular). The heritabilities of traits within each functional domain turned out to be more similar than between functional domains. Across all traits, heritability was 49 percent. For nearly 70 percent of traits, there was very little shared environmental effect, while genetic effects were almost entirely additive (see above). This study indicates not only that strong additive genetic components exist for almost all traits in the many publications analyzed, but that there is also an equally strong environmental component.

As any parent knows, the way a child is raised has profound impact; and, like all other kids, Deborah Skinner was doubtless influenced in ways unguessed at the time. For instance, the baby tender more than likely altered the microbial surroundings of its inhabitants. We now know that we live in a fantastically complex microbial world that impacts many of our most basic biological functions, and it is clear that the microbial component of any baby's surroundings is crucial to the development of its immune system. What's more, we now also know that the environment has an impact not only on how genes are expressed, but also on the ways in which genes are inherited. And as a result of our growing ability to observe how our environments alter the structures of our genomes, and how these alterations are passed on to the next generation, the new science of epigenetics has hugely expanded since the mid-2000s.

In an epigenetic response, the actual sequences of the genes do not change. Rather, the chromosomal structures around certain genes are altered by influences that ultimately emanate from some environmental factor, resulting in chemical changes that are passed along to the next generation. Studies of identical twins elegantly demonstrated this phenomenon, as researchers capitalized on the fact that MZ twins raised in different environments can be compared to determine where in their genomes those epigenetic alterations reside. To the great surprise of the scientists who initially investigated this phenomenon, pairs of identical twins raised together have typically shown very different epigenetic patterns of alteration on their chromosomes. We will have much to say later about the impact of environment on various human traits, but one thing to remember at this point is that environmental influences evidently tend to smooth out the distributions of traits in populations.

WHAT DID DARWIN DO RIGHT?

During his life Charles Darwin published twenty-five books on subjects ranging from evolution and coral reefs, to molds and orchids. He also ruminated at length on human variation and behavior, and obliquely referred to human origins. His famous notebooks also provide a firsthand view of his ability to describe objects and patterns in nature, and give us a window into his vast intellect. A century and a half later, Darwin's ability to synthesize information about the natural world remains almost without parallel. Yet it is the very nature of science to be a progress report that is constantly modified and refined, and subsequent evolutionary biologists have built upon Darwin's framework to create a very different picture of biology from the one that prevailed in the mid-nineteenth century when he lived and wrote. So, what is truly lasting in Darwin's ideas?

One of the very first things that Darwin did right was to develop a coherent and detailed argument for evolution, the notion that change occurs in the biota over time. Many nonbiologists mistakenly think that Darwin was the first to articulate the idea of evolution, but that is not the case; it was an idea whose time had clearly come even before Darwin's great work *On the Origin of Species* appeared in 1859. Indeed, Charles's own grandfather Erasmus Darwin had hinted at evolution in a long and semipoetic medical treatise called *Zoonomia* that he published in 1794. Not long thereafter, the French natural historian Jean-Baptiste Lamarck more practically documented change over time in lineages of fossils from the Paris Basin; and in 1814 the Italian paleontologist Giambattista Brocchi followed with an analysis of fossils from the foothills of the Apennine mountain range that suggested to him that species themselves, much like individuals, had beginnings, lifespans (during which they gave rise to offspring), and deaths (when they disappeared from the fossil record). Darwin himself was galvanized into writing the *On the Origin* when, in early 1858, he received a manuscript from a younger naturalist, Alfred Russel Wallace, that contained what he himself described as a précis of his own ideas.

But what Darwin did very clearly do, in a way that had never been done before, was to exhaustively document the evidence that ineluctably led to the notion that life had evolved. His attention to detail was stunning, and his observations of natural history were exquisite. And in his studies he came up with an explanation for the great organizing principle that natural historians and others had noticed among the denizens of the living world from time immemorial. Namely, that living things naturally group together into increasingly inclusive categories. For example, we human beings are most similar to the great apes; together, we most closely resemble monkeys; the larger group that includes monkeys, apes, and ourselves is most similar to the lemurs, with which we classify ourselves as primates; primates and similar clusters of furry beasts group into the mammals; the mammals belong, with other backboned animals, among the vertebrates, and so on up until, along with every other living organism in the world, we are subsumed into the megadomain Biota.

This is why every species on the planet can be represented as a tiny twig on a single great tree of life, those twigs coming together into ever-larger branches that converge toward a root at the base. And it was Darwin's (and Wallace's) genius to recognize that this structure of life is due to common ancestry. All life is descended from a single ancestor which, much as in a modern human genealogy, gave rise to multiple descendants, who themselves gave rise to descendants, and so forth. We resemble the apes most closely because we share a very recent common ancestor with them, while our common ancestor with all other primates lived earlier in time and the vertebrate common ancestor lived even longer ago. Hence Darwin's neat capsule definition of evolution as "descent with modification," which captures life's innate tendency to diversify. And indeed, Darwin's is still the only formulation we have that actually predicts the structure we see in nature.

Darwin also came up with an explanation for how evolution occurs. For all his brilliant originality of thought, the unfortunate Lamarck has been pilloried by history for concluding that the time-related change he saw among his fossils had been caused by the accumulation of novelties that organisms had acquired over their individual lifetimes (though with the emergence of epigenetics he has lately been enjoying a bit of a vogue). But Darwin came up with a magnificently reductionist explanation for change that helped him to sell his controversial ideas to a society that conventionally believed in the biblical creation story. And to put his explanation in context, let's quickly look at how natural history had conventionally been viewed.

Almost any consideration of how humans have traditionally studied nature needs to go back to Aristotle. In the Greek philosopher's *Parts of Animals* and *History of Animals*, we see a fairly sophisticated view of the living world, but one that takes a very specific and rigid stance. Known to philosophers as essentialism, this view suggested that organisms have "essences" that allow humans to identify them. This theory encapsulated a strictly typological way of viewing the biota. Typological thinking involves detailed examination of the natural world, but in the end the salient observations are boiled down to those that are helpful in identifying or "typing" organisms. This way of thinking is still a mainstay of modern taxonomic biology, in the sense that taxonomists use typology in their species descriptions and to produce the keys that many readers will have used in the field to identify unknown organisms. A quick look at any key—for plants or insects, say—is sufficient to demonstrate the typological nature of any taxonomic system, as you narrow down the identity of your unknown organism by following a chain of fixed attributes. Sometimes nontypological approaches are also used in modern taxonomy for exploring species existence or divergence; but in the long run the goal of taxonomy is to find the diagnostic characters that circumscribe species and larger groups. Up until Darwin, natural history had been more or less typological throughout, making the average, or the diagnostic, the key to describing nature. To achieve this, all variation was stripped away; and while it might get a mention in species descriptions, variation was usually treated as a byproduct that burdened the typological approach.

Darwin turned the typological world on its head. The process began during his long voyage around the world, from Britain to South America, the Galapágos and Australia, then back to South America and Britain, on the navy sloop HMS *Beagle* between 1831 and 1836. For reading on this five-year voyage Darwin didn't opt for light fare like Percy Bysshe Shelley's *Prometheus Unbound* or one of Sir Walter Scott's period pieces. Rather, he was deeply impressed by a book by the geologist Charles Lyell. Lyell was a leading advocate of Uniformitarianism, the dictum that all geological phenomena had resulted from the same small, incremental forces that can still be observed modifying the landscape today. Darwin learned from him that changes can be infinitesimally small, and he incorporated this idea into his thinking about evolution.

Lyell's influence can be detected in Darwin's ultimate decision that evolution boils down to the accumulation of small changes over large periods of time. But when the young naturalist returned to England in 1836, he also read *An Essay on the Principle of Population*, a 1798 pamphlet by Thomas Robert Malthus in which the demographer argued that resources limited the sizes of populations (and that human population growth imperiled progress toward a perfect society). With its emphasis on the merciless truth that not all members of a population could survive, Malthus's essay deeply affected Darwin's intellectual journey toward the idea that evolutionary change was inevitably driven by what he called "natural selection." Despite its rather deterministic name, natural selection was in fact no more than differential reproduction, whereby individuals with more favorable heritable characteristics survived and reproduced more successfully than those with inferior hereditary adaptations. Consequently, those superior adaptations were preferentially passed along to future generations, each of which would thereby differ slightly from the one before as the population's adaptation to the prevailing environment improved.

Not only did natural selection provide a mechanism for gradual change in the physical appearance of populations, but it placed those populations—and the variation within them—front and center in biology. After all, future improvements in each lineage of organisms would come from the variation already present among the individuals comprising them. And replacing tired old typology with "population thinking" meant that it was the average that became the abstract mental construct, while attention became focused on the variation around it.

In graphic terms, the distribution of a trait in an idealized population took the form of a symmetrical curve with a big hump in the middle and tailing off toward each end. We will see later how the striking shape of this frequency curve is critical to describing any population, because most variation among living organisms is distributed in this way, diminishing away from the mean. For any characteristic (weight, for example) there is an average population value. At the extremes of the curve, on either end, reside the individuals with the most extreme deviations from the mean (in this case, the lightest and heaviest). Typological thinking ignores this variation, and places all the emphasis on the mean. To a typologist, the variation away from the mean is an encumbrance that is basically to be ignored. In contrast, in populational thinking the mean is merely a statistical abstraction.

ROLL THE DICE

To describe a population in this way is to invoke the famous bell curve, or what statisticians call the "normal distribution." The bell curve derives its name from the shape of the curve as it proceeds from clustered values near the mean toward fewer individuals at the extremes. To explain this curve further compels us to invoke some statistics; but we hope that framing our discussion with some of the more amusing aspects of the history of this branch of mathematics will make the process a little more palatable.

Antoine Gombaud, Chevalier de Méré, liked to gamble, preferably using dice. Conventionally, a single die was rolled, and in that game de Méré knew that the probability of rolling any particular number (say, a one) was one in six. Bored with the simple game, he introduced a variant using two dice, and he set twenty-four as the number of rolls for an interesting bet (winnable, he thought, two out of three times, based on the correct assumption that two ones in two rolls are one-sixth as likely as a one is in a single roll). To keep the bet similar to the single-die game, de Méré simply multiplied the number of rolls by four, to get twenty-four. His reasoning was that he would roll two ones in twenty-four rolls as frequently as he would get a one in four rolls. In which case, his bet should not only be winnable, it would still have the substantial two-in-three house advantage.

To his horror, de Méré found that he was continually losing money on this more complicated game. But rather than quit gambling, he asked his mathematician friend Blaise Pascal (who consulted with another mathematician, Pierre de Fermat) why he was losing so much of the time. As gifted mathematicians, Pascal and de Fermat easily showed that the real probability of rolling a one twice in twenty-four rolls was actually only 0.49. The math of this calculation is not important here, but it did convincingly show that what de Méré had actually done in creating his new game was to shift the house advantage to his adversaries. It turned out that, while the game was exciting and close-run, those taking de Méré's bet would have a 2 percent advantage in the long run, which was why de Méré was losing his shirt. Helpfully, the mathematicians suggested that a small change, increasing to twenty-five rolls, would tip the odds slightly de Méré's way, since the probability of rolling two dice twenty-five times and getting a double one is 0.51.

For at least two millennia—and probably a lot longer—people like de Méré have enjoyed flipping coins and rolling dice. Not only can these activities provide hours of amusement to members of a species that is conspicuously intolerant of boredom, but to those like de Méré, there is something utterly compelling about the chance, even a risky one, of gaining wealth without effort. Naturally enough, the more ingenious gamblers have always sought to improve the odds of winning; as a result, it turns out that the raffish pastime of gambling has been the source of huge advances in the fields of probability theory and statistics.

What Pascal and de Fermat really did, in helping de Méré figure out why his betting luck had turned bad, was to estimate the very first binomial distribution. As the name suggests, binomial distributions are based on two-outcome systems. De Méré's problem had been that he was considering whether he would win or lose one dice toss at a time; and what Pascal and de Fermat did for him was to create a system whereby exact probabilities could be accurately calculated for successions of tosses. As it happens, a lot of information in nature is also binomial in character, which among other things makes binomial distributions important in understanding the data that scientists use to describe human populations. In addition, as we will see, estimating the binomial distribution for multiple categories of events led to the discovery of the "normal" distribution, discussed in greater detail below.

NORMAL DISTRIBUTIONS

Believe it or not, several treatises have been written over the past four centuries on the subject of coin flipping. One of the most interesting of them came from the pen of a French-born mathematician who fled Paris during the late-seventeenth-century persecution of the Huguenots and ended up in London. Abraham de Moivre was a brilliant mathematician to whom even the normally ungracious Isaac Newton frequently deferred, as one who "knows all these things better than I do." The pair were good friends; Newton considered de Moivre "one of the chosen few," and finagled a position for him on a committee formed to help decide whether Newton or Gottfried Leibniz had actually invented calculus. To Newton, de Moivre was an intellectual equal to such an extent that the two engaged in a scientific "show me yours and I'll show you mine" by exchanging controversial manuscripts before publication. Then and now, there is nothing more trusting in all of science than to share a manuscript or an idea before it is published. It is the scientific equivalent of teenage girls sharing diaries or chefs exchanging jealously guarded recipes.

De Moivre acquired a legendary reputation as a mathematician, and there is even a story that he predicted the date of his death through mathematical reasoning. It went like this. Toward the end of his life he became more and more lethargic; this required that he increase his sleeping time by fifteen minutes per night. Given this progression, he reasoned that at the point by which he was sleeping twenty-four hours a day, he would be dead. He accordingly died on November 27, 1754. While the legend is difficult to confirm, it does point to de Moivre's ability to understand number series. In fact, de Moivre is responsible for solving one of the more important number series in all of mathematics, the one known as "n!" or "en factorial." While this is a simple term to solve for small numbers ($1! = 1$; $2! = 2 \times 1 = 2$; $3! = 3 \times 2 \times 1 = 6$; $4! = 4 \times 3 \times 2 \times 1 = 24$; and so on), solving it for larger values of "n" is quite difficult.

Unlike his compatriot de Méré, to de Moivre games of chance and chess were simple entertainment. But on a more practical level, he earned part of his living advising gamblers. He was so adept at explaining the vagaries of coin tossing and other games of chance that his manual on the subject, a book called *The Doctrine of Chances: A Method of Calculating the Probabilities of Events in Play*, was published in four editions and in both Latin and English. During his excursions into games of chance, de Moivre recognized something very distinctive and unique about the distribution of results in coin-flipping games.

Gamers had recognized from the beginning that if they were flipping a fair coin (not all coins are fair, as anyone cheated by a lopsided coin can attest) 100 times, they would generally get fifty heads and fifty tails. They recognized that a fair coin meant that they had an equal chance of calling it right or wrong. In other words, they had a 50 percent chance of flipping heads, and a 50 percent chance of flipping tails. Winning on a bet to flip heads was simply a matter of having Lady Luck on your side. Now consider this. If you flip a coin 100 times, what is the chance of getting seventy or more heads? The probability is obviously less than 50 percent, but just what is it? At first, the solution appears simple, requiring you just to compute the chance of getting seventy heads. Right? Wrong! You also need to compute the chance of getting seventy-one heads, and seventy-two heads, and so on up to 100 heads. Computing this is a daunting and highly repetitive process.

To solve this problem, de Moivre did the following coin-flipping experiment. He calculated the chance of getting no heads, one head, and two heads in two coin flips. The chances are easy to visualize and then to graph, as we do below. Let's do the easy calculation first. The chance of flipping one head in two flips is 0.5, or one-half. Now for the tougher ones—no heads, and two heads. The probability of flipping no heads is the same as the probability of flipping no tails, which is also the probability of flipping two heads. The chance of flipping zero heads plus the chance of flipping two heads is equal to 0.5. So, if the chance of flipping zero heads and two heads is equal, this means that each has a chance of 0.25. This can be graphed as shown below for N=2, where N stands for the number of heads that appear, and the probability of getting each of these outcomes is given on the y-axis. Using the binomial distribution, we can do the same kind of graphing exercise for two flips, four flips, and twelve flips. For two flips, the curve looks relatively unspectacular and rough, and it doesn't improve much with four flips (N=4).

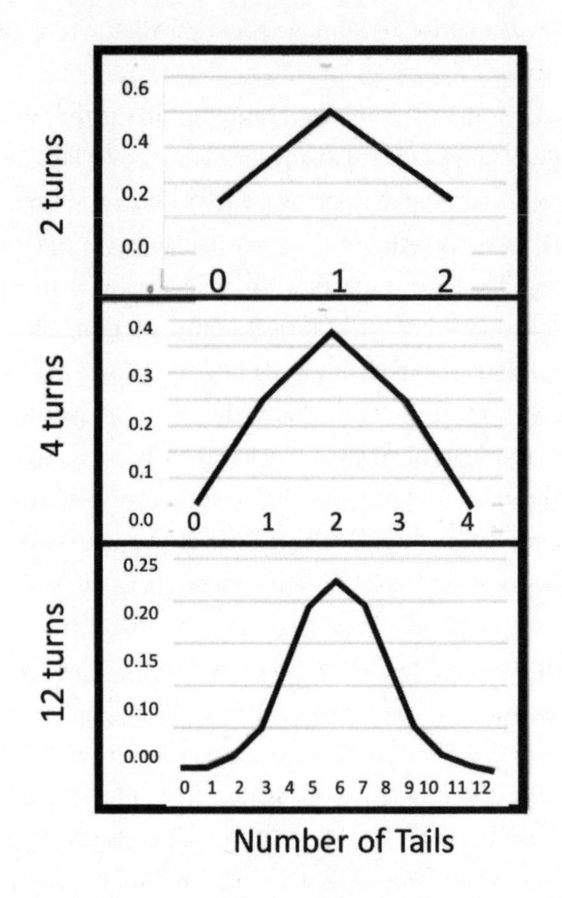

Results of the coin-flipping experiment, for N=2, N=4, and N=12.

But de Moivre realized that the more flips you do, the smoother the distribution will get. In fact, by the time you consider twelve flips (N=12), the curve is already relatively smooth. And what's more, it assumes the classic bell shape we are so interested in. So why is this curve so important? Well, it has some rather interesting properties that allow us to characterize sets of observations or groups of data in very straightforward and meaningful ways. For instance, as we saw earlier, a mean can be calculated from any set of observations, and for certain scientific questions, the mean is all that is necessary. But graphs like this also show the degree of variation from the mean that is so important in populational thinking. De Moivre's achievement here was to develop a normal approximation to the binomial distribution. Not all observations fit the normal approximation, but a great many do; in particular, many biological measurements of populations fit the bell curve or are at least approximated by it. Why should this be important? Well, it makes the math of doing comparisons much easier, it contextualizes everything much better, and, ultimately, it explains a lot about human behavior.

In the early nineteenth century, the mathematicians Pierre-Simon Laplace and Carl Friedrich Gauss showed that the normal distribution has some very interesting statistical properties. As a result, the curve became known as a Laplace or a Gaussian curve, and de Moivre almost lost out on the glory for discovering the darn thing. However, in 1893 Karl Pearson, one of the founders of modern statistics, suggested that the etymologically disputed curve simply be called a "normal" curve. He may have eventually regretted this, because in 1920 he wrote: "Many years ago [in 1893] I called the Laplace-Gaussian curve the normal curve, which name, while it avoids the international question of priority, has the disadvantage of leading people to believe that all other distributions of frequency are in one sense or another abnormal."

WHAT CAN A NORMAL DISTRIBUTION TELL US?

Not long after de Moivre proposed his curve, scientists started using normal distributions in their work. Astronomers were the first, but since this book deals with questions of human biology we should turn to Adolphe Quetelet, a nineteenth-century social scientist who used the principles set out by de Moivre to study chest girth in Scottish soldiers. Quetelet scoured the literature and found several papers in which chest measurements of Scottish soldiers were recorded. The figure below shows the table of his original data. (It turns out he made several mistakes in compiling the table, but the overall shape of the curve was still approximately normal when it was finally corrected).

Next to the table are the graphed girth data for this group of Scottish soldiers. Quetelet's take on this curve, though, was a bizarre reversion to typology, for he suggested that it visualized "the average man." That was because Quetelet assumed that there was a fixed direction in what nature produced. In essence, he supposed that nature was looking for a specific optimum that best allowed humans to be successful in life, and that any extreme phenotype was hence undesirable. In other words, he committed the error of thinking that, via evolution, nature strives for perfection. The reality, however, is to the contrary. What succeeds in nature is usually no more than the best available solution to an environmental challenge, given the variation that is at hand. And the best solution does not have to be anywhere near perfection. You can be the second-slowest member of your group and still avoid the predator.

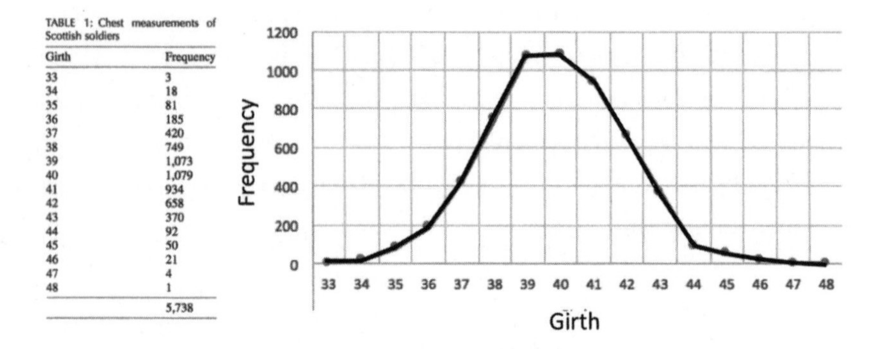

TABLE 1: Chest measurements of Scottish soldiers

Girth	Frequency
33	3
34	18
35	81
36	185
37	420
38	749
39	1,073
40	1,079
41	934
42	658
43	370
44	92
45	50
46	21
47	4
48	1
	5,738

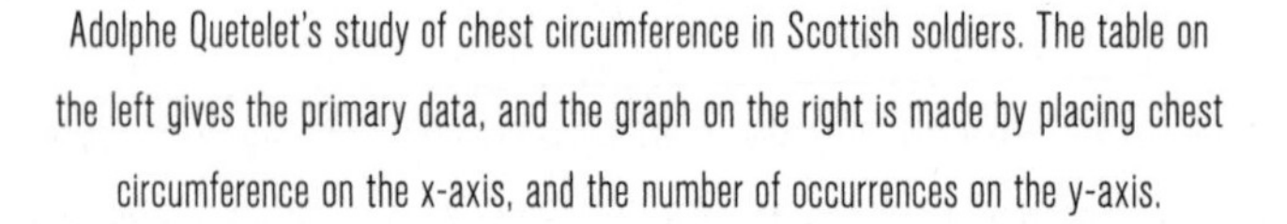

Adolphe Quetelet's study of chest circumference in Scottish soldiers. The table on the left gives the primary data, and the graph on the right is made by placing chest circumference on the x-axis, and the number of occurrences on the y-axis.

In 1914, Alfred Blakeslee produced a photograph (on page 33) of University of Connecticut students (at that time all males) arranged by height. The distribution here is not quite as good as in Quetelet's study (note the missing five-foot-ten and five-foot-eleven intervals), but it is readily recognizable as a normal distribution. Since one of the authors of this book is actually five feet ten but always claims to be six feet tall, we posit that this departure from the nicely symmetrical bell curve can be attributed to a little bit of fibbing by the students. The mean then becomes an important component of what the normal distribution can tell us.

Alfred Blakeslee's living histogram of University of Connecticut undergraduate men. Note that the categories at five-foot-ten and five-foot-eleven slightly mar the normal distribution.

If we look closely at a normal distribution, and we understand the axes well, then we can see another—more significant—aspect of the normal curve. On the x-axis are placed the values observed for a given feature. These values are collected by examining a population (of anything) for a specific quantifiable trait or aspect. In doing so, we might see one value occur many times. Hence, the y-axis shows the frequency with which the observations on the x-axis are observed. Note that in a normal distribution the "fat" part of the curve has the most observations for the given values. The tails of the curve represent observations that are made very few times. The distribution around the mean number of occurrences diminishes laterally, and it is fairly symmetrical.

But we aren't finished yet with the qualities of the normal curve. The two normally distributed curves in the figure on page 34 are both normal, and both have the same mean. But they look very different. What makes them different is how much variation there is in the data from which they are calculated. The black curve has many more categories of observations, and hence that curve is "fatter." The light gray curve is skinnier because it has fewer categories of observations. As a result, the black curve is said to have more variance associated with the mean. One of the interesting things that mathematicians who followed de Moivre did was to characterize this variance. Using the mean as a starting point, they were able to work out math that allows us to estimate the probability of finding a specific observation within any set of observations. The major breakthrough came with the definition of what is called the standard deviation of the mean, and the ability to compute it.

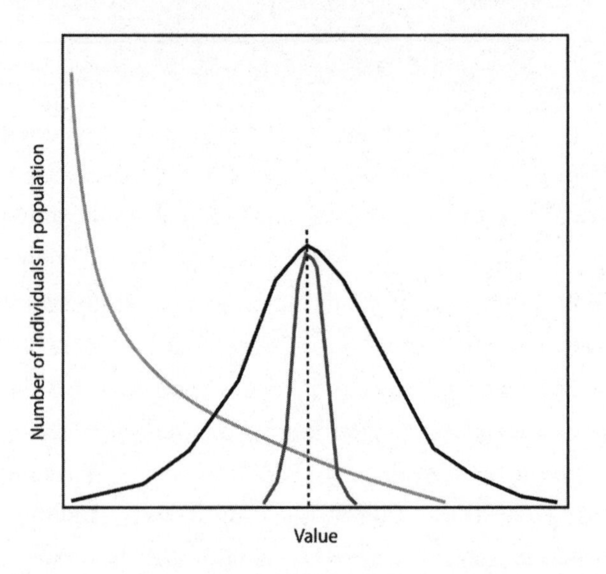

A comparison of two normal distributions, a skinny one (dark gray line) and a fat one (black line), with a Paretian distribution (light gray line).

The standard deviation allows us to estimate the probability that we will make a certain observation in a particular data set. So, for instance, an observation that lies within one standard deviation on either side of the mean will be made 68 percent of the time. An observation that lies within two standard deviations away from the mean will be made about 95 percent of the time, and so on, with increasingly large probabilities for greater standard deviations.

Let's return to our fat and skinny curves for a moment. The mean of both data sets is the same—let's call it 1.0. The standard deviation of the fat (black) curve is greater than that of the skinny (gray) curve. Let's call them

0.3 and 0.1, respectively. For the fat curve, an observation of between 0.7 and 1.3 will be made 68 percent of the time, and an observation between 0.4 and 1.6 will be made 95 percent of the time. For the skinny curve, though, 68 percent of the observations should lie between 0.9 and 1.1, and an observation will fall between 0.8 and 1.2 some 95 percent of the time. This property makes the standard deviation a handy tool for predicting what kind of observations will be made in any normally distributed data set.

BACK TO THE GENES

For present purposes it is important to know whether traits in human populations are normally distributed. Happily, even though some polygenic traits may not be distributed normally, most of them are because they tend to follow what is known as the "central limit theorem." This is a mathematical statement of the obvious. It requires us both to have a lot of observations (say, a global human population) and that those observations be independent of each other. If the observations show a well-defined mean and variance, they will be distributed normally. And if we return for a moment to our discussion of coin flipping, we will see a nice example of the central limit theorem in action. Not to be confused with the law of large numbers—which simply states that, given more and more observations, the mean of a sample will approach the mean of the population, making the estimation of the mean of a set of observations a snap—this theorem allows us to smooth out the distribution of phenotypes. Indeed, we would not be discussing normal distributions in connection with human traits if they did not fit the central limit theorem.

We can now begin to ask what traits in human populations are distributed normally. The best examples include height (i.e., the University of Connecticut undergraduate example from the early twentieth century), weight, body fat index (BFI), IQ, many human behavioral traits, blood glucose and triglyceride levels, muscle strength, muscle mass, depressive episodes, and many others. But we need to be careful, because if a trait does not fit a normal distribution, it will not have all those nice statistical characteristics. Thus, researchers from a business school recently examined the distributions of certain human activities involving individual performance. They examined five individual performance traits in 198 separate samples, with over 600,000 individual data points. "Traits" included the number of publications by science researchers, the number of Emmy nominations of entertainers, the number of times individuals were elected to the US House of Representatives, the total of NBA career points, and the number of errors made by MLB baseball players. They then tested whether the data for these traits were a better fit either to a normal distribution or to another kind of distribution, called Paretian. In most cases (around 90 percent) the statistical tests showed that individual performance data, at least in the five data sets just listed, fit the Paretian distribution better than a normal distribution. So, what is a Paretian distribution, and do results like this suggest that we need to reformulate our thinking about traits in human populations?

As seen in the figure on page 34, the Paretian distribution (light gray) looks like a ski slope. You still graph the range of responses on the x-axis, and frequency of observation on the y-axis, but the categories are changed to fit the Paretian distribution. Specifically, what is graphed here is a "transformed" version of performance (on the x-axis) versus the frequency with which you see such performance (on the y-axis). In comparing the normal and Paretian distributions, you can easily see the main difference between the two: both curves use the same raw data, but they use them in different ways. In a normal distribution, the majority of observations are at the mean of the distribution, or in the "fat" part of the curve. In the Paretian distribution most of the observations are at the left end (top of the ski slope), and there are very few observations made at the other extreme of the distribution. If the x-axis shows increase in individual performance from left to right, then this means that the distribution has very few superstars (high performers at the bottom of the ski slope). However, at the left side of the curve (top of the slope) there is a large category of individuals who are mediocre performers at best. The curve shows that the top 10 percent of performers comprise about 10 percent of the population, with middle performers comprising the next 10 percent, and, sadly, the other 80 percent of us at the bottom. This distribution fits the 80/20 rule first articulated by Vilfredo Pareto, the Italian economist for whom the curve is named. Pareto recognized that in Italy, 80 percent of the country's wealth was owned by only 20 percent of the people. This small percentage of Italians, it seemed, were financial superstars, and the majority of the population was poor in comparison. Now it should be evident why people in business schools should be interested, because the implication of the study is that individual business performance follows this 80/20 rule, with a small proportion of high-performing employees and a large number of underachievers.

Still, the fact that some traits—those in areas of human experience that have a few winners and a large preponderance of losers—may best fit a Paretian distribution will have little impact on our argument in this book. That's because there is something artificial in the Paretian distribution: it defines an extreme of behavior, and then categorizes everyone by how far away they fall from that extreme. In effect, the two sides of the normal distribution are combined into one, receding together to one side of the mean. If we were in an impish mood we might suggest that, instead of taking the human resources approach and graphing the range from low performance to superstars from left to right on the x-axis, we might do better to flip the logic of the Paretian curve on its head and graph a low propensity for bizarre behavior at the top of the slope, and a high propensity for such behaviors at the bottom of the slope, on the right of the graph.

One very basic fact shines through all this. Namely, that just as most physical traits are polygenic, most of the behavioral features in our vocabulary are far from discrete and most definitely polygenic. Some of them may seem discrete as you fume about the antisocial behavior and parentage of a driver who has just cut you off, but the fact is that the antisocial behavior itself is not. This is because, however it may be distributed, virtually any human behavior you can name actually lies on a spectrum: one that will correspond in one way or another to the normal distribution. Saintly or evil may be accurate descriptors of some individuals, but when you look at the human population as a whole, those apparently opposite conditions emerge simply as the two ends of the same curve. Most of the individuals from whose behaviors that curve is calculated will be reasonably congenial, and thus will lie somewhere in the middle—though where exactly they will fall on the curve will almost certainly shift a little from one day (or even one moment) to the next. The nastier and nicer people get, the farther away they will lie from the mean, more or less in proportion to their propensities. There are very few saints among us, and thankfully very few monsters, either. For pretty much any pair of behavioral antitheses you will find a similar bell curve joining them—and virtually any behavioral descriptor you can imagine has its antithesis.

CHANCE AND EVOLUTION

In the mature Darwin's view of things, evolution was a gradual process driven by natural selection. And while his central insight—that all living things are related by common ancestry—was accepted with surprising rapidity in stuffy Victorian England, his chosen mechanism for change remained a subject of lively controversy, even after the birth of modern genetics at the turn of the twentieth century. But during the 1920s and 1930s, coincident with the rise of quantitative genetics, a neo-Darwinian consensus known as the "new evolutionary synthesis" began to emerge. This brought natural selection back to the fore in evolutionary biology. Although the new way of looking at things came relatively late to paleoanthropology, once students of human evolution had taken it on board they did so wholeheartedly. From about 1950 onwards, virtually all English-speaking paleoanthropologists were happy to accept the idea that human evolution had consisted for the most part of the gradual modification of a single central lineage of hominids, via natural selection. As we will see in some detail in Chapter 3, we are still struggling with the fallout from this extreme viewpoint today.

Meanwhile, though, it is enough to point out that since the mid-twentieth century ample evidence has accumulated that the evolutionary process (or processes) is much more complex than the simple fine-tuning of lineages by natural selection. For one thing, it turns out that past climates and environments were highly unstable, changing on timescales that natural selection, by its nature a very gradual business, could never have kept up with. For another, a closer look at the fossil record shows that—as Giambattista Brocchi suggested back in 1814—fossil species have tended to linger more or less unchanged between their times of appearance and disappearance, rather than gradually evolving themselves out of existence as the synthesis suggested they should have done. And what is more, numerous new additions to the human fossil record have made it evident that, far from having been a story of the fine-tuning of one lineage, the saga of human evolution has been one of high drama as numerous new species (new variations on the human theme) were tossed out onto the ecological stage, to sink or swim in competition with each other and with other elements of the biota. What had appeared to be a slender evolutionary tree, in other words, turned out to look much more like an untidily branching bush. Clearly, a lot more was going on to promote change in the history of life than simple competition among individuals for reproductive success. Hominid evolution has very evidently been deeply affected by processes that unfurled at a much higher level than the individual: processes frequently involving environmental changes that occurred entirely at random with respect to the adaptations of the creatures affected.

The bottom line here, then, is that nature does not strive to produce perfection. Wherever you go in the world, you will see species competing with each other for ecological space in basically the same manner. But within species there are two possible dynamics going on as individuals compete for reproductive space. When a new mutation occurs in a large population, there are several possible outcomes. If the new mutation is disadvantageous, it will most likely disappear from the population. If the mutation confers a reproductive advantage, then it will in theory grow in frequency pretty much in proportion to the degree of that advantage—although we must remember that, because it is the entire individual that is the unit of reproduction, natural selection cannot work on individual traits. It works on populations of individuals; and it is, moreover, the whole organism, not the individual genes, that succeeds or fails in the reproductive stakes. This means that, however large or small the population, any gene has to propagate within a fantastically complicated and intricate genomic context. Finally, if a mutation's effect is basically neutral, it may simply hang around in the population, although in the end it will most likely be eliminated. As populations get smaller and smaller, however, the distributions of genes within them get odder and odder as a result of random sampling effects, or what geneticists call "sampling error." This produces a phenomenon known as genetic drift, which can radically change the genetic constitution of the population in a very short period of time. This is because in a small population a new mutation might rapidly rise to high frequency or even go to fixation (become the "new norm") simply as a matter of sampling error; or it might just as easily be lost for the same reason. Add to this the inherent randomness of mutation and of changes in the external environment, and you will readily see that evolution and engineering have little in common.

Nobody would dispute that organisms are in some sense adapted to their environments; but all adaptations are necessarily something of a compromise as a result of developmental contingency, because the kinds of structures that can evolve are constrained by the kinds of structures that already exist. Thus, with the possible exception of tumbleweeds, multicellular organisms cannot evolve wheels, because the developmental blueprints simply aren't there. Nonetheless, the blueprints that do exist hold the potential for all kinds of future changes, should they adventitiously happen. There is, indeed, a good argument to be made that all genetic and phenotypic innovations come into existence as "exaptations" (or "preadaptations"). These are randomly modified forms of existing structures that are available to be recruited by their possessors for new purposes. A favorite analogy, made famous by the paleontologist Stephen Jay Gould and the geneticist Richard Lewontin, is to the decorated spandrels typical of domed cathedrals. These triangular spaces frame some of the world's loveliest paintings; but they are actually structural rather than aesthetic features, necessitated by the pillars that support the domes. The decoration of these preadapted spaces was an afterthought.

You can easily see where we are going here. An improved understanding of the complexity of the processes propelling evolutionary change inevitably introduced chance into the evolutionary equation. Indeed, by now it appears that fortuitous factors have been a lot more important than natural selection as drivers of innovation in the natural world. There is no doubt that natural selection has an important influence on the history of species—after all, selection is a mathematical certainty in a world in which, as Malthus realized, more individuals are born than ever survive to reproduce. But its effect is far more often to promote stability as both ends are trimmed off those bell curves, than it is to nudge lineages in a consistent adaptive direction over vast periods of geological time. And this, of course, is why we have spent so much time talking about de Méré, de Moivre, and the others. Evolution is not a process of optimization to anything; instead, it is more a matter of simply getting by in a difficult and, more importantly, uncertain world.

TWO

SCIENCE AND BEHAVIOR: TRAPPED BETWEEN SIMPLICITY AND COMPLEXITY

Because *Homo sapiens* is in essence a storytelling species, it is understandable that even scientists sometimes yield to the siren call of reductionism. But we can't ignore the fact that, when behavioral scientists yield in this way, they trap themselves in oversimplified narratives both about how our species got to be the way it is, and about how it behaves now. This tendency to oversimplify is particularly problematic in evolutionary biology, because simplistic explanations have proven especially well-adapted to genetic (and more recently genomic) reasoning. A good case in point is the notion that such rarefied qualities as our ethical and political behaviors are quantifiable traits that, thanks to our newfound ability to trawl the human genome, we can examine for their underlying genetic or neurobiological bases. The idea here is that if we can find the genes that specify these behaviors, we will know how the traits are produced. And if we know how they are produced, we will also be able to explain why we have them.

As members of a (sometimes) introspective species, we cannot readily pass up any opportunity to improve our self-knowledge; and our new technological tools do indeed open up exciting potential prospects for understanding the mechanisms that underpin who we are. But in any area as sensitive as our human behaviors—and our ethical behaviors in particular—we can hardly ignore the obvious pitfalls if we get our explanations wrong. And that, of course, means that we can't start jumping to conclusions without first rigorously examining the assumptions on which those conclusions are based—a caveat that has to be at the fore whenever we try to make associations between genes and behaviors.

Perhaps the most familiar, and certainly the most compelling, of the reductive approaches to human behavior is the one often known as evolutionary psychology. This crossover discipline is based, among other things, on the premise that behavioral traits are "atomizable," meaning that they can be studied as independent entities. This is a necessary premise because associations between cause and effect can be made only when the effects (specific behaviors) have been isolated. Only then can the genes be lauded or blamed for those human behaviors, which can then in turn be explained in terms of inheritance from earlier ancestors. By following this path, some scientists have—usually controversially—decided that the genetic bases for sexual preference, religiousness, ethical behavior, intelligence, and many other very complex behaviors do not simply involve responses by our very complex brains to immediate stimuli, but are instead expressions of ancestral tendencies.

Evolutionary psychology is among those domains of evolutionary research that have sidestepped the need for specifically mapping traits in a genetic context, concluding that it is enough simply to know that our behavioral repertoire has a genetic component. And it is also a good example of why we need to be careful when we interpret human behavioral traits in evolutionary terms. The discipline sprang out of the sociobiology craze that began in the 1970s when E. O. Wilson, an expert in social insects, published his text *Sociobiology*. The premise of this volume was that behaviors are inherited, and thus can be studied like physical traits. Since the social dynamics of insects are hugely different from those among primates, it was hardly surprising that sociobiological explanations for human traits—especially behavioral ones—were widely disputed and controversial from the beginning. Still, Wilson's approach to understanding human behavior and evolution soon morphed quietly into evolutionary psychology, and it has gained momentum in this guise ever since, despite its many conceptual problems. A major reason for this success is that evolutionary psychology is a supremely reductionist approach to explaining humans' often bizarre behaviors, and the human mind—that storyteller's mind which loves to make simple connections—seems naturally drawn to reductionist explanations.

One premise of modern genetics is that if we understand the genome, we will ultimately understand the biology of everything around us. But while the genomes of organisms do indeed carry the basic information necessary for the construction of every living thing, the promise that we will find the exact genetic bases of our many complex traits has rarely been kept. Instead, we are usually left with a rather incomplete—albeit tantalizing—picture of how things *might* work in generating the trait we are interested in. And how something might work is simply a hypothesis. Science requires that hypotheses be tested—indeed, it is the ability to be tested that makes a hypothesis scientific. But there is a big difference between proposing a hypothesis and actually testing it. In the real world it remains true that, as the comparative anatomist and evolutionist Thomas Henry Huxley remarked a century and a half ago, the great tragedy of science is a beautiful theory slain by an ugly fact. Everybody loves beauty, but few of us want to participate in a tragedy; and in consequence it is often tempting for researchers to jump the gun, and to reach conclusions on the basis of beautiful hypotheses that have not been rigorously tested, or perhaps not tested at all. One of our colleagues, who shall go unnamed, once said: "Never let data get in the way of a beautiful hypothesis." While our friendships are still intact, we have to point out that data are what science is all about.

GRIST FOR THE MILL

Some scientists have been aware of the challenge posed by reductionism in evolutionary genetics and medicine for at least the last four decades. Among them is the distinguished evolutionary geneticist Richard Lewontin, who during the 1970s and 1980s published a series of papers on population genetics in which he outlined the role of genes and allele frequencies in the study of evolutionary biology. Lewontin did this from a position of authority because, together with Harry Harris and John Hubby, he had spearheaded the use of allozyme research in evolutionary genetics. This technological innovation started a revolution in the field of molecular biology because allozymes, molecules that can be easily located in the blood or hemolymph of organisms, are not only a direct downstream product of the genes, but also can be detected using relatively simple methods.

When gene sequence changes, it may or may not alter the genetic code of the gene involved, and consequently it may or may not alter the amino acid sequence of proteins produced by the gene. If the protein sequence *is* changed, the alteration sometimes results in a change in the net electrical charge of the protein produced. That change can be detected by putting the protein into a gel matrix, and then forcing an electrical current through the matrix. If the net difference in charge is positive, the protein will migrate through the gel matrix in the direction of its negative pole. If it is negative, the protein will migrate toward the positively charged part of the gel matrix. This property of proteins allowed Lewontin and other researchers to discover allozyme variants that are related directly to the actual sequence of the gene involved: a spectacular advance for its time, and Lewontin realized it was a complete game-changer when it came to appreciating variation in natural populations.

To understand why, we need to look at Lewontin's training, and at his particular take on genetics in biology. He studied at Columbia University under Theodosius Dobzhansky, a wonderful naturalist and entomologist who came from Russia to the United States in 1927. Dobzhansky worked in Columbia's famous "fly lab" under the tutelage of Thomas Hunt Morgan, one of the founders of modern genetics; and, after joining the faculty, he ultimately became the most important of the three major moving forces behind the evolutionary synthesis, the view of evolutionary biology, already mentioned, that was developed in the first half of the twentieth century. Along with the ornithologist Ernst Mayr and the paleontologist G. G. Simpson (both worked nearby at the American Museum of Natural History), Dobzhansky attempted to meld Darwin's ideas about evolution with the theoretical numerical work of another big three, this time the quantitative geneticists R. A. Fisher, J. B. S. Haldane, and Sewall Wright. These mathematically inclined biologists had between them developed an extensive quantitative approach to understanding how genes might evolve in populations. But it wasn't until their theories were connected to empirical observations from experimental genetics, systematics, and paleontology that a true synthesis could be accomplished; and in this case, data did not get in the way of a beautiful hypothesis. Lewontin was a beneficiary of this process of integration; but he knew that nothing in nature is simple, and in his influential 1974 book *The Genetic Basis of Evolutionary Change*, he summarized the situation as he saw it then:

> [Population genetics] was like a complex and exquisite machine, designed to process a raw material that no one had succeeded in

> mining. Occasionally some unusually clever or lucky prospector would come upon a natural outcrop of high-grade ore, and part of the machinery would be started to prove to its backers that it really would work. But for the most part the machine was left to the engineers, forever tinkering, forever making improvements, in anticipation of the day when it would be called upon to carry out full production.

Still, while he was critical of his predecessors (those "lucky prospectors"), Lewontin realized that what they had been doing was creating a robust framework for studying variation among and between populations of organisms. He then promoted his allozyme data in this context by describing them as new grist for the great evolutionary genetics machine that the early twentieth century had produced. But he soon began to detect a grinding of gears in this modified machine, and in 1991 he walked back on his admiration of allozymes by switching his metaphor for these genetic markers from "milestone" to "millstone."

Genomics had not yet fully taken hold at this point, but in a prescient discussion of the use of DNA sequences in evolutionary genetics, Lewontin rather coyly concluded that DNA "is another story, and anyway it is somewhat reminiscent of one that I remember telling before, about 25 years ago." This statement strongly implied that, in its turn, DNA also had the potential to become a millstone if inappropriately employed. Lewontin's long career at Harvard has extended into the genomic age, and he remains active well into his eighties. But lately he has been relatively quiet about the role of genomics in modern evolutionary biology, quite likely because, to his ear, that grinding of the gears he had detected earlier failed to change much when genomic data, rather than allozymes, began to be fed into the machine.

Over his long career, Lewontin intermittently reviewed the state of the art in population genetics. The last such review, published in 2002, discussed the future of evolutionary genetics and took the form of a sobering look at three undertakings that are basic to evolutionary genomics. First, Lewontin pointed out that evolution requires variation among individuals in a population, and that the first thing any good evolutionary biologist needs to do is to survey the system he or she is interested in for such variation, thereby also providing the raw data for any systematic study. The second task lies in the realm of comparative biology, because genome-scale data can give researchers a new view of what is shared both within and among groups of organisms. And the final endeavor involves the straightforward use of genomes as a means of locating individual genes. This last endeavor is largely a technological challenge, and it involves narrowing the limits on the kinds of genes that might be involved in determining phenotypes of interest. Many biologists have taken up this cause, to which we now turn.

WHAT ARE "GENES FOR"?

If you have recently opened a newspaper, fired up your computer, or turned on your television, it's a pretty good bet that you've encountered a report of a gene being discovered "for" this trait or that. But as attractive as they might be to journalists, such revelations are usually optimistic at best; and they are most often downright misleading. To substantiate this claim, we might do worse than to offer in evidence Richard Dawkins, the famous atheist and evolutionary biologist, who recently made a splash by discussing the evolutionary dynamics of homosexuality online. He repeatedly referred to "the gay gene" and to the "gene for homosexuality" which, by his account, persisted in human populations for a whole slew of reasons.

Some people (and some giraffes, for that matter) are homosexual. But what did Dawkins mean by the "gene for"? Well, his ability to visualize a "gene for X" (in this case, homosexuality) comes from his strong belief that it is the gene that is the unit of evolution, a notion that he cleverly, if unpersuasively, advocated in his now four-decades-old bestseller *The*

Selfish Gene. Dawkins defined the gene for X in a binary context, in which individuals either have the gene or don't. In Dawkins's view, a gene for X is any bit of DNA, however small, whose possessors have a higher probability of expressing X in their phenotype than individuals without the snippet of DNA do. But this actually gets us nowhere, because this definition is so broad as to be meaningless. That's because it would include not only any gene with a major phenotypic effect—sickle-cell disease, say—but also any gene that had even the most trivial of effects. Yet, clearly, only those genes that exert a major effect on the phenotype come even close to being satisfactory candidate "genes for X." And in the real world there turns out to be only a handful of genes of this kind, most of them behaving in a Mendelian fashion—although it's important to remember that, while by definition all Mendelian genes are also major-effect genes, not all major-effect genes are Mendelian. More importantly, though, most genes are pleiotropic—that is, they influence more than one trait—while most character spectra are polygenic, meaning that they are influenced by many genes. Seeking one-to-one correspondences between protein-coding genes and specific phenotypic characteristics is thus for the most part a fool's errand.

Understanding precisely how genotypes specify phenotypes, and how they influence such phenomena as speciation, have long been twin Holy Grails in evolutionary biology. The discoveries of such phenomena as the rules of inheritance, the linearity of genes on chromosomes, the process of mutation, the fact that DNA is the hereditary element, and genome sequencing were all big steps along the long road toward finding them. But, just like pinpointing the mythic vessel from the Last Supper, identifying and validating the exact genomic bases of phenotypes and speciation have in most cases proven extremely problematic, despite all advances. This is because, to repeat, the vast majority of phenotypes, both physical and behavioral, are determined in a complex polygenic manner, so that while a handful of traits (the simple ones) can meet the causality criterion, the vast majority almost certainly can neither be broken down into discrete units, nor be attributed to discrete causes.

One way to describe the genetic, genomic, and developmental complexities of many traits is to look at their genetic architectures. The term *architecture* accurately reflects this issue of complexity, because buildings can neatly be broken down into their architectural characteristics. Major categories might include roof, building materials, doors, windows, chimney, and so forth. Each of these categories will exist in a series of variants, and by summing up the variants desired, the architect can easily characterize the makeup, or the architectural character, of a particular design. The next step is then to encode the overall character of the house in a blueprint, so the builder can construct the house as desired.

The blueprint is the first step in allowing the builder to understand the architecture of a house, and this is where the analogy with complex phenotypes comes in. Trudy F. C. Mackay, a geneticist who studies genes and traits in *Drosophila* fruit flies, has remarked that "understanding the genetic architecture of quantitative traits begins with identifying the genes regulating these traits." But while the endeavor she describes might seem straightforward, in painful reality understanding the relationship between most traits and phenotypes is one of the more difficult goals in modern biology, certainly in the majority of cases in which the trait under consideration is complex. Still, this has hardly deterred speculation. The usual approach has first involved figuring out just where in the lengthy genome the genetically varying trait is controlled. The resulting genome-level information can then, in theory, be used to work out exactly which alleles are involved. But simple as this procedure may sound, it has often involved pulling off some remarkable technological feats; and even then, it is no sure thing.

QUADRANTS AND COMPLEXITY

If we have two populations with very different behaviors we can, in principle, easily compare the genomes of the two populations to identify where DNA variations responsible for the difference might occur. But reality is rarely that simple. For while there are two polar possibilities—that the genomic variation in question is highly correlated to the trait of interest, or alternatively that it is completely neutral—there will usually be a huge range of gene variants that have varying degrees of effect on the phenotype. When you are looking at genes or traits of major effect, picking out the correlated genes is not that tricky. Several genetic disorders, such as PKU (phenylketonuria, which we will return to later) and sickle-cell disease, actually had their genetic architectures determined before genomics came along. But in anything of lesser effect, it is often extremely difficult to discern the details of genetic architecture.

It might be a bit counterintuitive to turn to disease as a model for understanding the genetic basis of normal human physical traits and behaviors, but with respect to genetic architecture many diseases can actually be considered to be a lot like physical traits. And because it is easier to obtain funding for studies of disease than of evolutionary phenomena, a vastly larger sum has been spent on trying to unravel the genomic causes of clinical conditions than those of normal behavioral traits. The data set on clinical disorders is, accordingly, both huge and relevant to trying to understand the general rules underlying the genetic architecture of behavioral and other traits. Let's take a closer look.

One important factor here is known as "the spectrum of the alleles." This describes the characteristics of all the alleles involved in a disease, among them the number and locations of the places in the genome at which DNA sequence variation impacts the expression of the disease. Also important are the frequencies of the disease alleles in the population at large. As we have learned more and more about human genomes and allelic variation, it has become obvious that allele variants involved in disease are rare compared to other kinds of variants in the genome. What's more, most diseases turn out to involve many allelic variants, so that each allele will typically have a weak effect on the overall expression of the disease. Indeed, on its own, a single disease allele might not be detrimental at all. As we know, variation in our genomes arises by mutation, a process that is continually occurring.

Mutations are point events that can easily be lost during the reproductive process and that can increase in frequency in populations only very slowly. As a result, most detrimental variants will eventually be lost, and those we do see are most likely due to recent mutations. What this means for the allelic spectrum of disease is that most of it will consist of rare variants that have minimal effect on the disease trait.

In considering the various kinds of genomic changes that will impact evolutionarily important traits, or ones thought to be involved in our behavior, it is useful to borrow from the medical genomics literature a concept called "common variant-common disease" (CV-CD). Originally put forward by David Reich and Eric Lander in 2001, this concept was intended to guide clinical geneticists in dealing with genetic disorders in the context of genomics. Reich and Lander recognized that human genetic disorders could be characterized along a continuum, or spectrum. And they pointed to two variables that geneticists had been using since the dawn of the genomics age in medicine. The first of them, which we've already mentioned, concerns the effect of the gene or genes involved, while the second concerns the frequency with which the disease alleles occur in human populations. Having two variables like this means that they can be visualized in two-dimensional graphic form, as in the figure below. In one dimension we have allele effect, also known as penetrance, which is just the degree to which an allele combination affects the interactions that give rise to a particular phenotype. On the other axis we have allele frequency, the measure of how rare or common an allele is. This way of mapping complex traits conveniently divides the graph space into four quadrants. In the first quadrant, at upper left, we have alleles with high penetrance and rare allele frequency. This quadrant holds most, if not all, of the Mendelian traits that human geneticists have mapped so far. Because such alleles are both rare and highly penetrant, they are easy for scientists to work with, albeit often devastating to the people suffering from them. If a nasty disease is highly penetrant or has a large effect, its genetics will be clear cut and its frequency should be quite low.

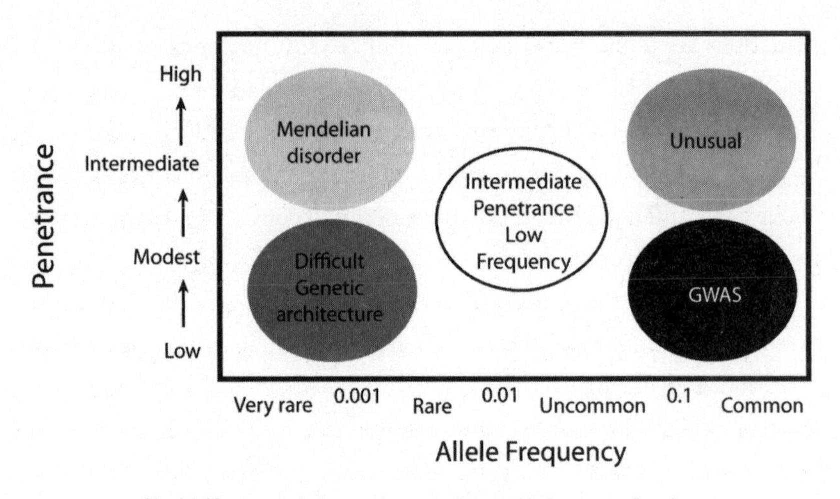

The CV-CD space with four quadrants, and a possible "sweet spot" in the middle. *Modified after McCarthy, et al. (2009), graphic by Kayla Younkin.*

The second quadrant holds alleles that are also highly penetrant, but that are found in high frequency. Human geneticists believe that there are very few disease alleles of this type in human populations because of the detrimental nature of any highly penetrant disease allele. However, for evolutionary studies in which a trait may actually be advantageous, this category could be a gold mine. And there are indeed some very neat examples of advantageous alleles in this quadrant, including pigmentation alleles and those that control scale characteristics in fish.

The third quadrant includes alleles that have low penetrance or effect, combined with high allele frequency. Alleles in this category are detected by the association studies we will shortly discuss (GWAS and QTN). The genetics of alleles that behave in this way require large samples of genomes to tease apart their genetic architecture. But while alleles in this quadrant are a major focus of genomics research, they account only for a very small percentage of the variation in any given trait.

The final quadrant holds alleles of low penetrance and low frequency. These are very hard to detect with current methods and available sample sizes. Some optimists suggest that eventually larger data sets and very sensitive statistical methods will detect such alleles, but for now they are rather elusive. In addition, they suffer from the same problem that the third-quadrant alleles have. Namely that, even when found, they will not account for much of the causation of the trait under study, and so won't be very helpful. Still, there might be a sweet spot, right in the middle of the graph, in which alleles are both of intermediate effect or penetrance, and of intermediate frequency. Association studies ought in principle to be quite effective in this sweet spot; and in addition the alleles here add up quickly, to explain a higher percentage of the variation in a trait even than in the bottom-right quadrant where association studies do their best work.

WHERE'S WALDO?

So how do we find those important genes that are at the heart of quantitative traits? Well, you might liken the search for genetic causation to the well-known children's book series *Where's Waldo?* These books are replete with complicated drawings in which the eponymous Waldo is hidden and must be found. Traits in quadrant 1 are like the second-simplest of the puzzles in the *Waldo* books, those in which it takes relatively little effort to identify Waldo in his red-striped shirt and his red-and-white beanie with a red pompom on top. We can see Waldo in this quadrant because his form and colors permeate the overall picture, and he is there frequently enough to pick him out with relative ease. The easier quadrant 2 puzzles show only a handful of figures. Waldo's form and colors penetrate the other information easily, and it takes no effort to find him. This would be too easy even for a kindergartner so, just like quadrant 2 in the CV-CD space, there aren't many of these puzzles around. Quadrant 3 is like a puzzle in which Waldo is rather well camouflaged and it takes some effort to locate him in the overall picture. His visual penetrance is low to the reader, but he is there in high enough frequency that careful study of the drawing reveals him sooner or later. Quadrant 4 is the most perplexing. This is like a Waldo drawing in which he is nicely camouflaged, and the presence of tens of thousands of other figures in the drawing make the process of elimination very difficult. Not surprisingly, there are very few, if any, Waldo pictures like this, and very few, if any, alleles that have been detected in quadrant 4.

Staying with the Waldo metaphor, we can also discuss the science behind detecting alleles in each of the quadrants. Quadrant 1 is rather trivial. Since Waldo's image readily penetrates through to your vision, the task at hand merely involves examining the rather large number of figures present in the drawing and recognizing which one is Waldo. Quadrant 2 is somewhat irrelevant too, as the problem is even easier here. Waldo jumps right out at you, so this category is no fun. In contrast, quadrant 4 is so difficult that only the most dedicated Waldo fans would even attempt to puzzle it out. As a result, it is in quadrant 3 that things get most interesting for Waldo enthusiasts. Waldo is sufficiently camouflaged to make finding him a challenge, and there are in fact only two ways to do it: luck, and a very logical stepwise approach that involves sectoring off the drawing and focusing on the subareas in sequence. The latter is hard work, but it's very rewarding when you finally do find Waldo. The same goes for Quadrant 3 in medical genetics.

In the days before genomes were available, genetics researchers used an approach called quantitative trait mapping. Any mapped gene for a particular species that could be shown to have a linkage with a quantitative trait was called a quantitative trait locus (QTL). Most QTLs were mapped in favorite model organisms, or in crop plants like corn. For instance, *Drosophila* geneticists like Trudy Mackay used the technique to explore the genetic basis of complex traits like the number of bristles the flies have on their bodies.

Now that modern genomics is on the scene and single nucleotide polymorphisms (SNPs) are used to analyze genomes, the QTLs have become transformed into quantitative trait nucleotides, or QTNs. QTN (and indeed QTL) analysis relies on a genomic phenomenon called linkage disequilibrium (LD). This term simply measures how tightly linked two items are to each other: in other words, how close they are to each other in the physical genome. The closer they are, the higher their LD. If two items are completely unlinked, then they are said to be in linkage equilibrium. The first step in any QTN study is to find individuals with a specific phenotypic manifestation corresponding to the genetic trait you want to map. Those individuals are most readily produced by controlled crosses, something that can be done very efficiently in model organisms like *Drosophila*, or even the mouse (*Mus musculus*). Then DNA is made and sequenced from these individuals, and LD is measured between the phenotype and the QTNs included in the study. The QTN with the highest LD is the genomic variant closest to the genetic locus associated with the phenotype of interest. In some cases, the QTN might even be *within* the genetic locus that is involved in the trait.

You will readily appreciate that there is a problem here for those interested in locating human behavioral traits, because it is ethically inadmissible to control or direct mating in humans. Human QTN analysis can, however, fall back on known familial relationships in the form of human pedigrees, working with serendipitous phenotypic variation in the pedigree. The inability to control crossing makes things a whole lot harder, but in principle human QTNs might work if the trait under consideration were of major effect. However, it remains possible that such QTNs or QTLs might actually not be points, but rather regions with a number of alleles that might be relevant to the phenotype; and in such cases the full genetic underpinnings of the phenotype might be missed. But while major-effect traits and phenotypes of this kind do exist, they are far from the norm. And yet again, it's often a long shot that QTNs will uncover any trait or phenotype genes at all. For all these reasons, some researchers think that QTN mapping will help with our explanations of phenotypes only in very special cases.

If the applicability of QTNs to humans is limited, what else can we do? Well, instead of using controlled crosses, researchers have cleverly fallen back on the fact that population history serendipitously shuffles the genetic elements to produce decipherable patterns of LD among loci in natural populations. By separating a large population into individuals with the desired trait (cases) and without it (controls), researchers can in theory efficiently scan genomes for the LD of specific phenotypes. When such LD is found an association is said to exist, giving rise to the technique's formal name: genome-wide association study (GWAS). We haven't discussed the statistics involved in this process; suffice it to say that there are critical statistical cutoffs that researchers use to differentiate among associations. Those statistics are based on the fact that occasionally an association will appear randomly, and they determine the level of departure from random.

In biology, GWAS is like a controversial politician in the public sphere. To some people, such studies are the best thing since sliced bread, whereas to others they are anathema. In politics, supporters usually highlight the successes of their candidate, and deflect the question whenever shortcomings are brought up. Opponents, on the other hand, go straight for the jugular and point to the politician's weaknesses, character flaws, and failed endeavors. Depending on whom you support, you might consider some of the political verbiage as "fake news." In politics nowadays, the only way to get past the fake news is by diligent fact-checking; and it is the same with GWAS. As evidence of success, GWAS supporters point to the huge number of associations that have been generated over the past decade. Here is an example from the GWAS Central website, where all GWAS studies in the literature are summarized: "GWAS Central contains 69,986,326 associations between 2,974,967 unique SNPs and 829 unique MeSH [medical subject headings] disease/phenotype descriptions." In other words, as of late July 2018 nearly seventy million associations had been discovered between three million variable nucleotide positions (SNPs). If we associate those SNPs with the 800 or so phenotypes that are listed in the Medical Subject Headings database, a "medical vocabulary" of mostly disease-related human phenotypes, we get an average of nearly 4,000 SNPs per phenotype.

So a lot of work has been done, but how useful is it really? One way to measure the effectiveness of any study examining a phenotype or trait is to ask how much of the variation of the trait can be explained by a single variable. Sometimes this measure can be very high. Many Mendelian traits, for example, have very high percentages (as high as 100 percent) of total trait variation explained by associated SNPs. But more commonly, GWAS trait studies show only very low levels of variation explained by associating traits to SNPs. Indeed, one study revealed that over 100,000 different SNPs show association with human stature. And New York University's Matt Rockman points out that, in a GWAS of blood pressure done in 2009 on nearly thirty thousand people, only 1 percent of the trait variation could be explained by the top ten strongest associations. This means that 99 percent of the variation in the trait could not be explained by the ten best genomic associations together. If a politician reported that level of success, he would be accused of incompetence—or, worse, of being excessively honest.

We can learn a lot about the limits of GWAS, and about our understanding of heritability, by taking as an example one relatively simple inherited disease. Phenylketonuria (PKU) is a genetic disorder in the PAH gene that is responsible for processing the amino acid phenylalanine. PAH codes for phenylalanine hydroxylase, an enzyme the breaks down phenylalanine. When the protein doesn't function properly, phenylalanine accumulates in tissues of the body and causes severe intellectual disabilities and heart problems. In US and European populations, the disease used to occur in about one in twenty-five thousand people; and it was traditionally considered a recessive disorder because a "normal" PAH allele on one of the two chromosomes in a genome was enough to permit the production of normal PAH, and thus the normal breakdown of phenylalanine.

But it turned out that PKU—like most genetic disorders traditionally considered Mendelian—is in reality a lot more complex genetically than our classical understanding of dominant versus recessive loci would suggest. In fact, there are over sixty distinct lesions in the PAH locus that will cause PKU if present in the homozygous form, or in combination with some other PAH-deficient allele (a genetic state known as pseudohomozygous). However, if individuals are either homozygous or pseudohomozygous, and their diets are altered to exclude phenylalanine, they will not express the symptoms of PKU. Accordingly, high protein foods such as red meat, fish, milk products, and aspartame (an artificial sweetener) are avoided in PKU diets. The modified diet individuals exhibit perfectly normal neural development, so PKU is now relatively rare in modern Western populations because of early diagnosis followed by a phenylalanine-avoidance strategy.

The lesson of PAH alleles here is that, to develop a complete understanding of genomic variation, we need to fully understand the phenotype itself, especially in the context of environment. In a population with many instances of PKU, and diets uniformly high in phenylalanine, a large amount of the variation of the disorder would be explained by genetics. Indeed, the amount of variation of PKU explained by genetics in these populations would be near 100 percent. But if a different population had a PKU-causing allele that was fixed (i.e., all individuals were homozygous for the PKU-causing allele), but had a phenylalanine-free diet, the expression of the phenotype would be 100 percent environmental, and hence 0 percent genetic. The genomicists Paolo Vineis and Neil Pearce looked closely at this hypothetical situation and eventually had to conclude that "genetic and environmental components are inseparable." Likewise, any measure of heritability of this trait in these two hypothetical populations would be substantially context-dependent.

The lesson is that one cannot define percentages of heritable variation without knowing something about both the environment and the interactions of genetic loci. This means that, in any quantitative analysis, some important information is invariably either lost or was missing in the first place. That poses a significant problem for GWAS and other quantitative molecular studies as indicators of genetic association. What's more, there may be other reasons for some of this missing heritability. Recognition of this weakness has led to a flurry of research, as a result of which it now seems likely that, in addition to the environmental aspects, the missing heritability comes from SNPs with effects that are simply too small to be detected by the statistical methods used in GWAS.

Advances in technology have meant that more and more data can be thrown at this problem. But can they solve it? According to several studies, the answer to this is no; we will inevitably end up without having localized a significant proportion of the variation. Also disturbing is the realization that many of the localized genetic SNPs or factors do not reside in protein-coding regions of the genome, but rather affect regulatory aspects of gene expression. These regulatory regions occur in parts of the chromosomes that are called "promoters" and "enhancers," whereas Mendelian diseases and QTLs are localized to protein coding regions of the genome.

Given all this, it increasingly appears that complex disorders—and, by extension, complex phenotypes, including behavioral ones—consist mainly of weak-effect SNPs driven by key genes and regulatory pathways. Referring back to our example of height, with which 100,000 SNPs can be shown to be associated, we might thus better extend our understanding of the genetic architecture of complex traits by adopting an "omnigenic" view, one in which a vast number of genes and genetic interactions are involved in determining the phenotype.

As it happens, this omnigenic perspective harks back to a century-old idea about quantitative phenotypes that was first articulated by R. A. Fisher, one of the founders of population genetics. In 1918 Fisher was confronted with two competing views of phenotypic control: one from the beanbag Mendelians, and the other from the "biometricians" who looked at continuous traits. At the time, Mendelian genetics had enjoyed a run of almost two decades (ever since the rediscovery of Mendel's laws in 1900) as the prevailing paradigm in understanding change. But there was clearly more going on out there in nature; and, while the biometricians did not contest Mendel's tenets where they appeared to apply, they were impatient to understand traits, such as height, that they felt simply did not behave in a Mendelian fashion.

Fisher resolved the issue by showing that, as Jonathan Pritchard and colleagues neatly put it:

> If many genes affect a trait, then the random sampling of alleles at each gene produces a continuous, normally distributed phenotype in the population. As the number of genes grows very large, the contribution of each gene becomes correspondingly smaller, leading in the limit to Fisher's famous "infinitesimal model."

And the really key thing here is that Fisher's infinitesimal model leads to that normal distribution of phenotypes: the bell curve.

A DIGRESSION INTO PURE LUCK

The geneticist Matt Rockman has remarked upon a subtle realignment of evolutionary perspectives in the decades after genome sequencing appeared on the scene. He suggested that, although the synthesis had viewed large-effect variants as atypical, research in the area eventually began to witness "a quiet realignment toward a view of readily discoverable large effect alleles as the primary molecular substrates for evolution." This realignment involved some very interesting work whereby "lucky prospectors" (to use Lewontin's phrase) came across large-effect phenotypes that had evolutionary implications so strong as to explain something fundamental about the process of biological change. The best examples of this judicious prospecting typically came about when the researchers had both a predetermined idea of the genes that might be involved in a phenotype, and some convincing information about the trait's evolutionary relevance. Genes with such relevance are called "candidate genes," and identifying them requires both cleverness and luck.

One of the best examples of a candidate gene that paid off perfectly involved coat pigmentation in rodents, and what is known as adaptive melanism. It has long been known that fur pigmentation provides an adaptive advantage to animals living in certain environments, and the poster rodent for this is the rock pocket mouse *Chaetodipus intermedius*, which lives in the southwest United States in environments where the landscape is usually a bright sandy white. But intermittent volcanic eruptions have extruded large black lava patches across the landscape in the rock pocket mouse's range. Researchers had noted that the rock pocket mice populations in the area consisted both of mice with dark black pelage, and mice with lighter sandy white pelage. Predation on mice is high, and in terms of camouflage it is obviously advantageous for them to be colored to match the landscape they are living in.

Because the genetics of pigmentation in mouse hairs had been worked out in detail, Michael Nachman, Hopi Hoekstra, and Susan D'Agostino were able to propose a candidate gene for this phenotype in nature. Pigmentation, it had been learned, is the product of the interaction of two protein products: the melanocortin-1-receptor (MC1R) and the agouti-signaling protein (ASP). Nachman and colleagues decided to look at these two candidates. By examining the DNA sequence variation in and around the MC1R and ASP genes in the rock pocket mouse genome, they discovered that variation in the MC1R gene can be directly correlated to the light/dark phenotype. Since this initial discovery, Hoekstra has analyzed the evolutionary dynamics of the MC1R and ASP genes in natural populations of the mice, observing some very interesting evolutionary phenomena along the way.

The rock pocket mouse story is a wonderful combination of insight, hard work, and luck; but things are not always so simple. As a budding graduate student, one of us was convinced that a simple major-effect system was at work in some of the Hawaiian fruit flies he was studying for his thesis. One group of those flies is named the "modified mouthparts group." This is an apt name, for the mouthparts of these flies are an anatomical mess. Typically, those mouthparts are simple extensions of the head plates that the flies use to soak up food (usually yeast), and they look rather round and spongy. But the modified mouthpart flies have all manner of strange appendages where the "sponge" should be. Why? Well, it turns out that the males of these species also use their odd mouthparts to clasp female genitalia during the courtship process, so that what they give up in efficient feeding they make up by assuring successful reproduction.

It also turns out that a mutant of the common lab fly *Drosophila melanogaster* (a relative of the Hawaiian flies) also grows appendages out of its mouthparts. The mutant is called proboscipedia, and is caused by simple alterations to the *D. melanogaster* proboscipedia (Pb) gene. A nice candidate gene indeed, for an important anatomical anomaly in nature! Sadly, though, after months of looking at sequences of Pb genes in the Hawaiian flies, the only possible conclusion was that the flies showed no variation in this particular gene—and with no variation, a simple major-effect genetic explanation just got up and flew away. It was a hard way to learn both that luck is undependable, and that most traits of evolutionary significance are polygenic, involving large numbers of genes that interact with each other in producing the phenotype.

HEIGHTS AND FACES

Although it is by now pretty evident that quantitative genetics and association studies can take us only just so far into dissecting genetic traits into their component genes, relating genes to traits remains a major goal in genetics. And there have been some semi-successes that we think are quite instructive. One of the most complex traits that has been examined in this way is height, a feature that has been studied for well over a century. Indeed, it was a particular favorite of Darwin's cousin Francis Galton, who discovered the statistical principle of correlation while he was investigating variation in height.

Long recognized as a classic example of the normal distribution, the human height phenotype has recently been subjected to GWAS by the appropriately named GIANT (Genetic Investigation of Anthropometric Traits), an international consortium of researchers. Prior to GIANT's most recent work, GWAS studies had detected a couple of hundred gene regions that could be associated with height. By expanding the sample of genomes examined to over a quarter million, and by using a statistical trick that lowered the threshold for recognizing an association, the GIANT researchers found over four hundred genes that were involved in the genetic architecture of height. And even more amazingly, those four hundred genes explained only around 10 percent of the physical variation in the trait. Clearly, a bewildering variety of genes (many of them more generally involved with growth) is involved in the determination of height, resulting in a complexity that will continue to defy simple explanation at the genetic level.

Another trait that has garnered a lot of attention is the human face. Human beings are hugely social creatures, and faces are immensely important in individual recognition and thus in the maintenance of social cohesion. True, individuals who suffer from prosopagnosia ("face blindness") get by pretty well using voice alone, but this hardly detracts from the importance of faces in our daily lives. Twin studies indicate that about 60 percent of face blindness can be explained by genes; but more importantly here, the face itself has proved to be a very tractable system for genomic studies. This is because it is a complicated structure with numerous "landmarks" (see figure on page 66) that can be recognized not only by the human brain, but by the artificial intelligence of computer facial-recognition systems. It is largely variability among those landmarks that make each face unique and distinctive. Several research groups have used GWAS to dissect the genetic architecture of the human face, and they have converged on the conclusion that facial shape is a polygenic trait under the control of about forty genes.

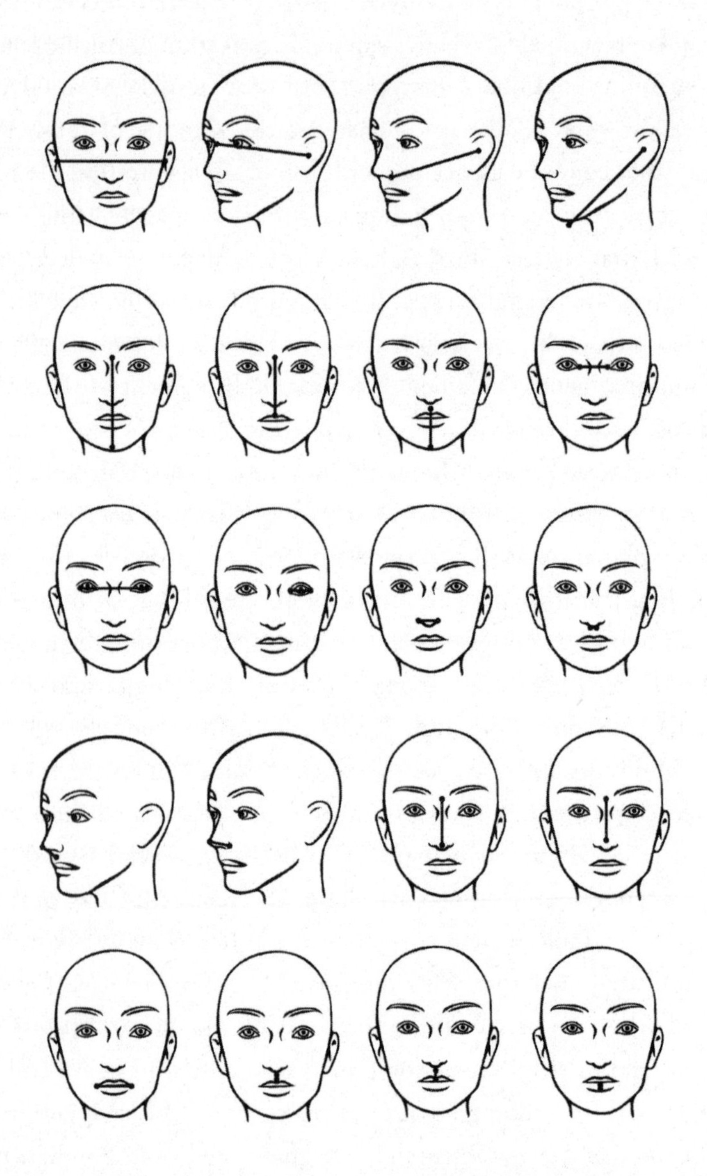

Landmarks used to define facial shape. The black lines indicate the facial measurements that can be used to characterize the landmarks. *Adapted from Shaffer, et al. (2016), graphic by Kayla Younkin.*

Since various facial landmarks have been associated with specific genes in this way, researchers have suggested that facial shapes can be predicted from genome sequences, something that has obvious forensic applications. But faces are not just identifiable by shape, and among distinctive non-shape features, eye color has been associated with eight genes. Skin color is even trickier: some forensic scientists use a panel of eight to ten SNPs to predict skin tones, but others suggest that this number is too low to be helpful in the context of the forty genes that have been associated with the trait. Oblivious to the practical complications Craig Venter, the man who competed with the National Institutes of Health to produce the first human genome, has used whole genome sequences to reconstruct human faces, claiming success in eight out of ten trials.

All this brings us to the key question of just what a face is, and to the key difficulty of setting parameters for anything as complicated as a face. Richard Spritz and colleagues, authors of one of the studies on facial appearance, suggest that "the human face is an array of variable physical features that together make each of us unique and distinguishable." But how distinguishable? How many times have you spotted what you thought was a friend or a celebrity in a restaurant, only to realize it was someone else? It takes only a few features for the human brain to join up the dots and make a connection. That's the basis for the doppelgänger phenomenon: a subset of the possible variation in a face is expressed, and voilà! A stranger's face we can relate to. So, do we call the facial phenotype the minimal set of landmarks we need to identify an individual face, or does the phenotype include whatever variables we can conjure up when we look at a face?

This problem of subtraits included in a larger trait gets even worse when we start to consider behavioral conditions, in which few well-defined landmarks are available. Normal individual human behaviors stubbornly resist quantification and may vary wildly from day to day. (For example, which side of the bed did you get out of this morning?) Some human psychological disorders have appeared to be better candidates than routine behaviors for genetic analysis, schizophrenia being among the most promising. But even the relatively clear-cut diagnostics of schizophrenia have so far defied significant genetic dissection. Clearly, when it comes to the matter of associating genes with behaviors—and particularly with those behaviors of interest to evolutionists—we are still on very shaky ground.

NORMS

If we go back to the CV-CD quadrant graph we looked at earlier, we can now better appreciate its relevance to evolutionary biology. Quadrants 1 and 2 hold Mendelian traits, and while these include some interesting conditions like sickle cell and adaptive melanism, effects in these quadrants are not very pertinent to understanding most phenotypes of potential evolutionary importance. It is in quadrants 3 and 4 that we find the most interesting phenotypes in evolutionary biology—and these are the ones controlled by alleles with small effects. What does this tell us about the traits that we might be interested in with respect to human populations and human evolution? Well, it tells us that these traits are either impossible to decipher, or that they require a case/control approach to map them. And since the case/control approach is at the very least controversial, the traits of greatest interest to evolutionists turn out to be the ones that we are least likely to be able to properly dissect genetically.

So enter the norm of reaction, a concept originally introduced by the botanists Jens Christian Clausen, David Keck, and William Heisey (hereafter CKH), who seventy years ago or thereabouts wrote several long papers describing their plant-transplanting work in the Sierra Nevada mountain range. The purpose of this research was to test the response of several local endemic plants to changing environments. By intensive breeding in the greenhouse, CKH were able to create several "clone-like" plant strains that they then planted at localities along a transect that ran from Stanford, California, up the western slope of the Sierra mountains, and then down the continental side of the range. Among many other things, CKH studied the response of genetically identical plants of the genus *Achillea* (commonly known as northern yarrow) to three different altitudes: at sea level (30 m at Stanford), at medium altitude (1,500 m at Mather), and at high altitude (3,000 m at Timberline). They used the height to which each plant grew as an indicator of its adaptation to the environment it found itself in. Because the plants were genetically identical, any variation among them from one altitude to another would best be explained by the environmental interaction with the genes. In the absence of any environmental impact, CKH predicted that no difference in height would be found.

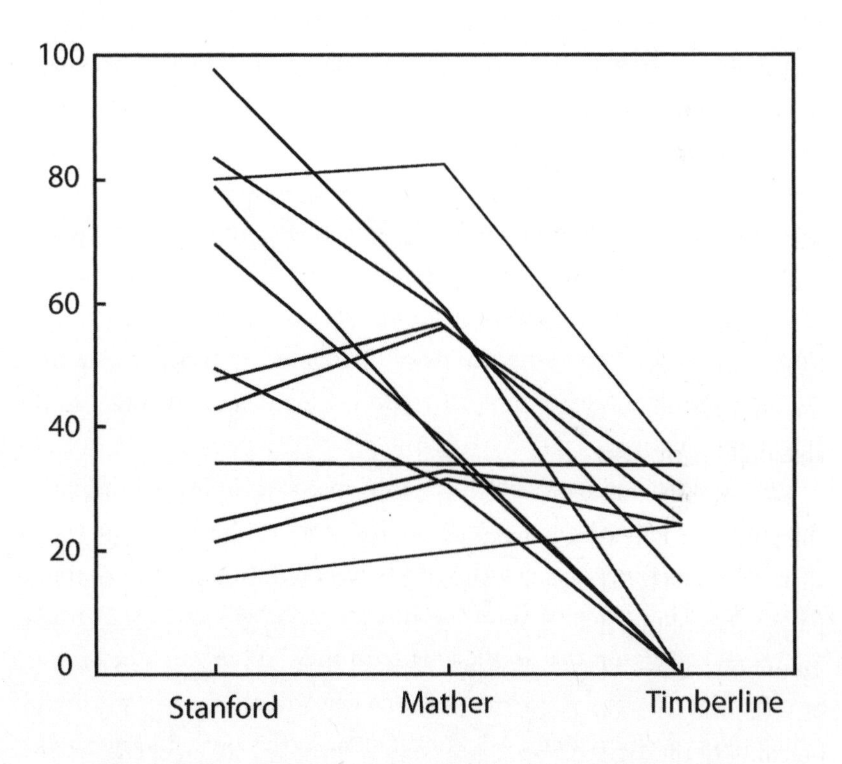

Reaction norms for maximum stem length of clones of various different populations of the yarrow, *Achillea*, at the three transplantation sites along a transect from the California coast to the Sierra Nevada. *Data from Clausen et al. (1948), graphic by Kayla Younkin.*

The figure shows CKH's results for clones made from plants taken from twelve original sites. The variation among them was remarkable. Some plants were oblivious to altitude. Some grew least at the lowest and highest altitudes, with the greatest growth occurring at midaltitude. Some showed a linear increase in growth, from lowest to middle to highest altitude, while others showed a linear decrease. Basically, all possible variations of response were generated by the experiment, demonstrating pretty conclusively that the environment an organism lives in—or creates for itself—is an important arbiter of its eventual phenotype. From these discoveries came the idea of a "norm of reaction," whereby the genotype specifies not an exact outcome to the developmental process, but instead a range of possible outcomes that the environment will select from.

Several GWAS studies have since been completed on the genetic architecture of height in plants. Most of these either used the model-organism plant *Arabidopsis* or crop plants, and the number of genes detected ranged from about thirty to over a hundred. A core of about seven genes was typically common from study to study, but the general indication of these association studies was that height has a complex genetic architecture. As usual, Lewontin summed it up beautifully: "a genotype does not specify a unique outcome of development; rather it specifies a norm of reaction, a pattern of different developmental outcomes in different environments."

Norms of reaction have now been shown to exist in many other plants and animals. Experimental and theoretical work in this area continues to be vibrant, although it is not as well funded as, say, research in genomics or GWAS. That is a pity, since the efficacy of GWAS and QTN studies is low, and we doubt that in the long term they will prove to be of much use in evolutionary studies—though one can still argue for some limited utility in the medical arena. We are not alone in this view, and researchers such as Matt Rockman and Kenneth Weiss have made similar arguments about the applicability of these genomic approaches to evolutionary studies. Both researchers seem to believe, as we do, that maybe we are using the wrong tools to answer key questions about complex traits. Or perhaps that we are asking the wrong questions. Rockman observes a "mismatch between question and method," while Weiss argues that we should strive for a "clearer goal." Fortunately, in the end we simply might not need to know the fully atomized genetic basis for traits to make progress in many areas of evolutionary biology. But in the context of understanding the genetic basis and heritability of human behaviors, we have reached a point at which we need, at the very least, to specifically acknowledge that the traits we are most interested in are complex, and thus are not susceptible to simplistic atomization.

One final note before we move on. It is possible to correlate almost anything with something else. For instance, there is a linear correlation between N (the number of firefighters) and $ (the dollar amount of damage of a fire). The correlation is mathematically a good one, but it suggests the illogical response of sending fewer firefighters to bigger fires. Failing to look beyond the superficial correlation might prevent you from realizing that fires that are bigger need more firefighters, because the real cause of damage is not the number of firefighters (N), but rather S (the square footage of the fire). The fallacy here is blindingly obvious, but in biology things are not always so clear-cut. Consequently, both the quantifiable elements of any problem, and exactly how they relate to causation, need to be made very clear if we are making claims about the history of traits in a population, particularly if they are behavioral ones.

Our survey here has made it pretty clear that, tempting as it may be to try, complex traits (the vast majority of heritable traits, both physical and behavioral) cannot be atomized for the purposes of inheritance studies. We must not only always bear in mind the norm of reaction, but also the fact that the genome is a hugely complex place. Virtually all physical and behavioral features fall on a spectrum; and, as Ronald Fisher showed a century ago, an infinitesimal model of causation invariably leads to the normal distribution of phenotypes. The inevitable corollary is that the behaviors we focus on in this book have to be interpreted not in terms of genetic determinism for individual points or extremes on the curve, but in in the context of the bell curve itself.

And this, above all, is why we need to exercise caution when we hear the evolutionary psychologists telling us that this or that human behavior—infidelity, for example, or violence—is an evolutionary holdover from times when such behaviors were more appropriate to the environment, the idea here being that our lives have changed so quickly in recent times that the slow mechanisms of evolution have simply not been able to catch up. For while it is certainly true both that human beings sometimes behave in bizarre ways, and that the environment of the overwhelming majority of humans has radically changed over the last ten thousand years or so, what is even more remarkable is how flexible and adaptable to changing circumstances people and their behaviors have proven to be. This flexibility underscores just how little members of our species are driven by inherited suites of genes to behave in the ways they do. After all, as far as the individual is concerned, infidelity is a choice—and it is not a choice that everyone makes.

As we will emphasize later, cultural accommodation has lately become a much more important and enduring force than biological adaptation in the molding of human experience. It remains true, of course, that we are nonetheless a variation on what went before, and that inside our heads there is still a lot of the ancient primate (indeed, vertebrate) brain. Accordingly, as any primatologist will readily affirm, many of the behaviors we see among our primate relatives—and particularly among the catarrhine primate group to which we belong—find strong echoes among human beings as well. Clearly, such comportments as infidelity and violence are far from unique to humans. But our recently acquired—and, more importantly, emergent—mechanisms of symbolic consciousness overlie and modify our behaviors in unique ways, and in doing so have changed all the behavioral rules by which humans play.

So why do many scientists succumb to the siren temptations of genetic reductionism? The obvious—and probably correct—answer is that even scientists may sometimes be uncomfortable with uncertainty, and that genetic and genomic hypotheses promise clear-cut cause-and-effect explanations when other avenues fail to provide them. Science frequently seeks to be predictive, and genetics is a very logical approach to understanding the world, replete as it is with rules, codes, and dogma. These features are not inimical to good science, because they are based on what philosophers of science would call background knowledge, which provides the essential connective tissue that holds scientific frameworks together. And with an inferred genetic basis, we can be more predictive than we could be otherwise about the outcome of a trait's interaction with the environment and about its evolutionary significance. But while all this added information gives researchers more confidence in making suggestions about any given trait's evolutionary validity or adaptive importance, it hardly gives them license to jump to conclusions simply on the basis of knowing that the basic blueprint for any individual resides in his or her DNA.

The second reason for capitulating to genetic reductionism is simple human hubris. When humans are told that something is very difficult or even impossible to do, they immediately attempt to do it anyway. And it is not just in science that we sometimes try to overplay our hand even when there is plenty of mathematical or theoretical evidence that a solution simply can't be found. This is not to say, of course, that a long shot may not sometimes work; and at this point you might be saying to yourself, "with thinking of this kind we would never have made it to the moon." But, as the great biologist Peter Medawar once declared in the title of one of his books, science is the "art of the soluble." And in the case of human behaviors, thinking within the constraints of what is possible should either make us realign our theories about gene associations, or back off and find another way to explain what we observe.

We hope that all this will not make us sound like Luddites. But it remains true that the machine that was built for detecting associations doesn't work as well as it should, certainly in an evolutionary context. And as a result, evolutionary biologists clearly need another mechanism for dealing with complex traits. Fortunately, one has been at hand for the last hundred years. Because what we do know with confidence about complex traits is that they fit the model, developed in 1918 by R. A. Fisher, which places them in the infinitesimal realm of a normal distribution: the bell curve. By settling on this distribution as an explanatory tool we can make better sense of our bizarre human behaviors, and of the equally bizarre ways in which they fit the norms of reaction.

THREE

EMERGENCE OF THE HUMAN COGNITIVE STYLE

Although human beings, logical thinkers at least some of the time, tend to think of themselves as very special creatures, there is no question that they come from humble beginnings. For, while modern humans undeniably possess some impressive cognitive specializations, this was not always the case among our ancestors. As members of a hugely egotistical species, we have not always found this mundane reality easy to accept; even the scientists tasked with interpreting the evidence for human biological history took some time to come to terms with it. Paleoanthropologists in the early twentieth century concluded, for example, that since it is our big brain that most strikingly distinguishes *Homo sapiens* today, it was this feature that should be sought in any fossil claimed to be that of a human forerunner. But the hugely embarrassing episode of Piltdown Man, the hoax fossil "discovered" in 1908 that combined the large braincase of a modern human with a piece of ape jaw, showed very clearly just how dangerous it can be to take a self-centered approach to defining who we are today and who we were in the past. The "brain first" Piltdown myth was definitively exploded in the early 1950s, and subsequent findings have shown very clearly that the human cognitive attributes we most prize today are not necessarily the features that set us on the path to becoming our modern selves.

ANCIENT APES

In sharp contrast to early conclusions that were based on sparse evidence and misrepresentation, today's large and rapidly expanding human fossil record suggests that the progenitor of our family Hominidae was a creature very unlike us. (Some prefer the subfamily Homininae; either way, it's the zoological group containing us and our immediate fossil relatives to the exclusion of our close living relatives, the great apes.) Almost certainly, that ancestor was a fairly run-of-the-mill member of the very diverse "ape" division of a larger grouping that also contains the monkeys of the Old and New Worlds. Ancient apes began to flourish in the forests of the Old World during the Miocene epoch, which spanned from about 23 to 5.3 million years ago. Indeed, those early times were the apes' heyday; their modern representatives—gorillas, chimpanzees, and bonobos in Africa, orangutans and gibbons in Asia—survive only in much diminished, and highly threatened, diversity.

Members of our forest-living progenitor species—not yet known in fossil form—probably differed from most of their ape contemporaries in being very highly suspensory in the trees, which meant that they would have kept their trunks quite upright as they moved around in their leafy habitat. Otherwise, in most respects they would have been pretty unremarkable compared to other apes. They evidently subsisted on a diet that consisted mainly of fruit and young leaves, processing it using both large chewing teeth housed in apelike projecting faces, and long, bulky intestines. Relative to body size, their brains would at best have been no bigger than those of modern apes. And while their cognitive systems were undoubtedly complex for their time (for example, they might well have been able to intuitively understand, as living apes evidently can, that other individuals might have harbored beliefs they knew to be false), they would have shown none of the specific cognitive peculiarities that are the hallmark of *Homo sapiens* today.

In Africa the Miocene was a period of dramatic change, both topographic and climatic. At the beginning of the epoch, tectonic forces lifted and cracked open the interior of the forest-clad continent, along the line of a current of molten rock rising from the interior of the planet. This upheaval created today's familiar East African Rift system that stretches from the Sinai to Mozambique. By blocking moisture-laden winds from the Atlantic, the upraised highlands and volcanic domes along the Rift placed the lands to the east in the rain shadow of the damper west, while the steady subsidence and filling of the valley floors as the Rift continued to develop led to the preservation of an astonishingly complete fossil record of the animals and plants of Africa over the last twenty-five million years.

By the middle of the Miocene worldwide climatic changes were also beginning to affect Africa, as the Antarctic ice cap expanded, the oceans grew cooler, and climates became more seasonal. At around sixteen million years ago, the consequent decline in summer rains in most of Africa was beginning to affect the nature of the forests, and hence of the primates that inhabited them, so that well before the end of the epoch dense forests had been widely replaced by vegetation adapted to seasonal rainfall regimes. This drying out led eventually to the appearance of the now-familiar open savannas and woodlands of eastern and southern Africa. The slow and step-wise fragmentation of the forests had profound consequences for our ancient Miocene and Pliocene progenitors, opening up new woodland and bushland habitats that those highly suspensory apes traversed on two legs rather than four.

THE EARLIEST HOMINIDS

Nobody knows for sure what a very early hominid ought in principle to look like, since as we go back in time toward a common ancestor the divergent lineages will show steadily diminishing differences. But as increasing numbers of putative ancient hominid fossils are found at African sites dating to between seven and four million years ago, paleoanthropologists are focusing their attention on two anatomical complexes that are well preserved in, or inferred from, the hard tissues that mineralize and form the fossil record. The first of these is the locomotor system. *Homo sapiens* is unique among living mammals in being a striding, erect-bodied, terrestrial biped. Our upright posture is reflected throughout our skeletons, from the way our flat-faced heads are balanced atop vertical, S-curved spines, to our short, stiff feet with their stubby and aligned toes. Our legs are long, our arms and hands are relatively short, and our pelvises are narrow and bowl-shaped to support the contents of the abdomen that now lie above them.

In all these respects, our structure contrasts strongly with that of habitual quadrupeds such as the living African apes that, for example, have long-faced heads that project in front of a normally horizontally held vertebral column that lies in a convex curve. This curve is matched below by the contour of the belly, and abdominal muscles that not only provide opposition for those of the spine, but also support the abdominal organs that are slung below it. The apes' arboreal propensities are, moreover, directly reflected in their relatively long arms, short legs, and in their slender, grasping hands and feet, with short thumbs and divergent great toes.

Some elements of the human bipedal locomotor system are present in every fossil contender for the title of earliest hominid. And the same is true for the main features of the other notable anatomical complex we mentioned, which resides in the dentition. The long-faced apes have large canine teeth that lock together when the mouth is closed, the upper canines coming to rest between the lower canines in front and the front lower premolars behind. These latter teeth are modified to provide long, forward-facing surfaces against which the dagger-like upper canines hone themselves as the teeth occlude. In contrast, modern humans have flat faces bearing notably small canine crowns. These mostly project little, if at all, beyond the rest of the toothrow; and the premolar honing surface is completely absent.

We are not entirely certain of the degree to which upright posture and canine reduction—both thematic in certain ape lineages, as well as among hominids—are directly functionally related. But what does seem to be pretty clear is that no arboreal ape descending to the ground as its ancestral forest habitat began to fragment would ever have chosen to locomote upright on terra firma unless it had instinctively found it the most natural thing to do. After all, in evolution each change is inevitably constrained by what went before, and upright posture would have only felt natural for a creature that already habitually held its trunk erect while moving around in the trees (as today's gibbons and sifakas do, primates that also travel upright on the ground in their own ways). To put the matter slightly differently, there is no plausible scenario whereby any committed arboreal quadruped would ever have decided to walk awkwardly upright on the ground simply because this made it possible for it to see farther in the newly opening landscape, or to free its hands to manipulate objects, or to carry things, or to thermoregulate better, or to attain any other of the various "key advantages" of terrestrial bipedality that have at one time or another been proposed.

What's more, once any ape had committed itself to walking upright on the ground, it would immediately have been the beneficiary of *all* the advantages of being a biped in this hazardous new milieu—and of all the *dis*advantages as well, notably slowness of movement and a vulnerability to knee and ankle damage.

Some fossil apes of the period between about thirteen and nine million years ago have bony features suggesting they had unusually erect arboreal postures, thereby hinting that multiple lineages of arboreal apes may have been experimenting in this direction back in the Miocene. But there is currently no material evidence to suggest the emergence of an anatomically distinctive and evolutionarily cohesive family Hominidae before about seven million years ago, just as the epoch was about to end. This is the age of a cranium (skull without lower jaw) discovered in 2002 in the central-western African country of Chad. Dubbed *Sahelanthropus tchadensis* by its French discoverers, the specimen is fractured and distorted. But a virtual computer reconstruction suggests that, although *Sahelanthropus* possessed a quite protrusive face with long tooth rows, the foramen magnum (the large hole through which the spinal cord exits the brain) pointed downward, rather than backward as a quadruped's does. The skull of *Sahelanthropus* had evidently been balanced on top of a vertically held spine, something that strongly suggests erect posture—and therefore bipedality—when the creature was on the ground.

In addition, although the preserved canine teeth of *Sahelanthropus* are quite pointy they do not project excessively beyond the toothrow, and they show no evidence of honing. The chewing teeth behind them are squarish and flattish, and thus not out of place for an early hominid. None of this guarantees that *Sahelanthropus* was directly related by descent to later hominids, because convergence—the independent acquisition of similar structures in different lineages—is not an uncommon evolutionary phenomenon. But the fact that the Chad fossil resembles us in both of those two key systems is quite compelling.

There is, alas, little more we can currently say about *Sahelanthropus*, except that the ecological preferences of the animals it lived alongside suggest that both forest and grassland existed nearby, and that the size of its brain, at a reconstructed 378 ml in volume, was comfortably within the ape bracket—our brains clock in at around 1,330 ml, compared to a chimpanzee's at around 400 ml.

Slightly younger than *Sahelanthropus*, at around six million years old, are some fossils from Kenya that have been given the name of *Orrorin tugenensis*. They consist mainly of some femora (thighbones) that, although broken, are plausibly those of a bipedal form. Some possibly associated teeth are also compatible with hominid status. Accordingly, despite the limited information we have about it, *Orrorin* appears to be a reasonably plausible claimant for membership in the group that contains the later hominids, and it helps confirm that hominids were on the landscape by six million years ago.

Later in time, and much more informative, is the 4.4 million-year-old *Ardipithecus ramidus* from Ethiopia, itself putatively preceded by various earlier Ethiopian fragments, named *Ardipithecus kadabba*, that are up to 5.8 million years old. The later *Ardipithecus* (see figure on page 82) is represented by a fragile and fragmentary skeleton that is nonetheless complete enough to give us a pretty good idea of the entire organism. At well over 50 kg (110 pounds), *Ardipithecus* was relatively large for an arboreal primate; it is also said to have various features, including a relatively stiff foot structure, that suggest it moved bipedally when on the ground. Still, in contrast to those typical of later hominids, the feet of *Ardipithecus* had divergent big toes that could grasp, as is typical of arboreal primates. As a denizen of a woodland habitat *Ardipithecus* plausibly spent a significant amount of time moving bipedally on the ground, suggesting that, as it were, the big toe may have been the last element of the hominid foot to have come into line. The brain of *A. ramidus* was a little smaller than that of *Sahelanthropus*, and it had a similarly protrusive face; but its chewing teeth were unspecialized, and its canines were fairly small and did not show much size difference between the sexes. All in all, then, the array of features we see in *Ardipithecus* is generally compatible with hominid affinities.

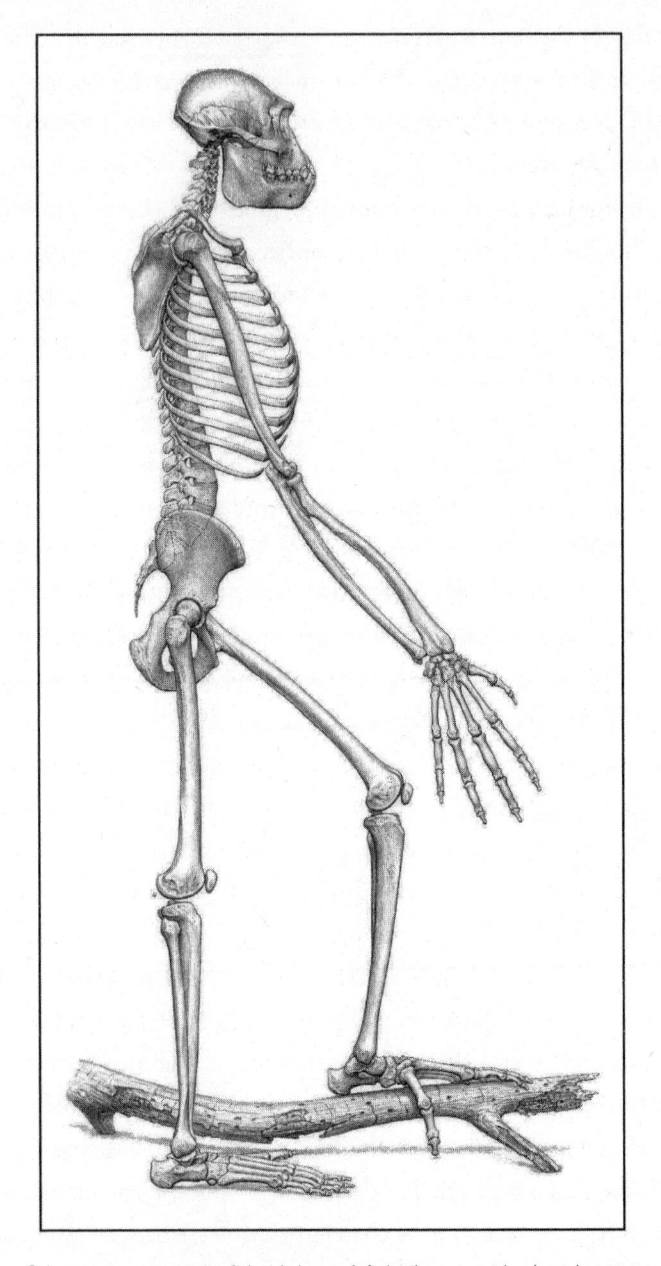

Side-view reconstruction of the skeleton of *Ardipithecus ramidus*, by paleoartist Jay Matternes. Note particularly the divergent great toe that contrasts with the in-line great toe of the modern foot. *Courtesy of Jay Matternes.*

Considering all three of these earliest hominid candidates together, one feature stands out. Namely that, aside from the common thread of bipedality, they make up a pretty motley assortment of fossil primates. Still, although it is perhaps unlikely that any of these pretenders actually stood directly in the modern human line, one or more of them may have been quite close to the hominid stem. And certainly, it is reasonable to conclude from the evidence they provide that the hominid family was already in existence by around seven million years ago. What's more, taken together with suggestions of erect posture in a couple of slightly earlier fossil ape species, these very early putative hominids strongly support the idea that, during the climatic changes of the latest Miocene, Mother Nature was experimenting quite vigorously with the evolutionary potential inherent in being a bipedal ape. This trial-and-error process established a pattern of significant diversity—and presumably competition—that subsequently held throughout the evolution of our family, almost up to the present day. Clearly, Hominidae has consistently been as diverse as any other successful mammal family. In and of itself that is hardly unusual. Other such families—Old World monkeys, for example, or the one containing cattle and antelopes—typically also have a diverse array of species and genera within them. What *is* truly unusual about the hominids today is that our species *Homo sapiens* is now the lone hominid in the world.

THE AUSTRALOPITHS

Much better documented than the early strays we have just looked at are the "australopiths" that lived between about 4 and 1.5 million years ago. First recognized in South Africa in 1924 following the discovery of a juvenile specimen of the now well-known species *Australopithecus africanus*, this group of early bipeds is now widely represented at numerous fossil localities in southern and eastern Africa, and even in Chad far to the west. The most ancient australopith species, *Australopithecus anamensis*, has been identified from teeth, jaws, and some elements of the body skeleton found at sites in Kenya and Ethiopia and that date from the period around four million years ago. Its jaws are long and parallel-sided, indicating a projecting face; but its canines are quite small, and there is no trace of premolar honing. Even more suggestively, the structure of the tibia (the large lower-leg bone) of *A. anamensis* shows definitive evidence of bipedality in both the knee and the ankle joints: as in humans, the femur (thighbone) angled in toward the midline as it descended from the hip joint to meet a tibia that rose vertically above the ankle. Poorly known as it is, the structure of *A. anamensis* strongly supports the notion that the hominid family got its start as a branch of the apes that had adopted bipedality on the ground as the ancestral forested habitat began to come under pressure. This agrees with what we know of the conditions in which it lived: based on the faunas associated with it, *A. anamensis* inhabited a mosaic of habitats that tended toward the closed, with plenty of trees around.

Traditionally, scholars have recognized two broadly defined australopith groups: the "robust" and the "gracile" versions, both known from sites in southern and eastern Africa. In general, it is reckoned that the last robust became extinct without issue around 1.5 million years ago, while the ancestry of our own genus *Homo* lay somewhere within the gracile clade, the latest member of which was still going strong at around the two-million-year mark (see the adjacent human family tree). The robusts possessed huge chewing teeth, coupled with greatly reduced canines and incisors, while the graciles showed the reverse, with relatively large teeth at the front of the mouth (though canine crowns were low) and relatively small premolar and molar teeth (though they were still pretty big by the standards of later hominids). Three or four species each of robusts and graciles are usually recognized, but without question the best-known and most emblematic australopith species is *Australopithecus afarensis*, to which the Ethiopian fossil skeleton dubbed "Lucy" famously belonged. The exact relationships of *A. afarensis* within the wider australopith group are a matter of debate; but we are probably justified in simplifying our discussion here by taking this species as an exemplar of the australopiths as a whole.

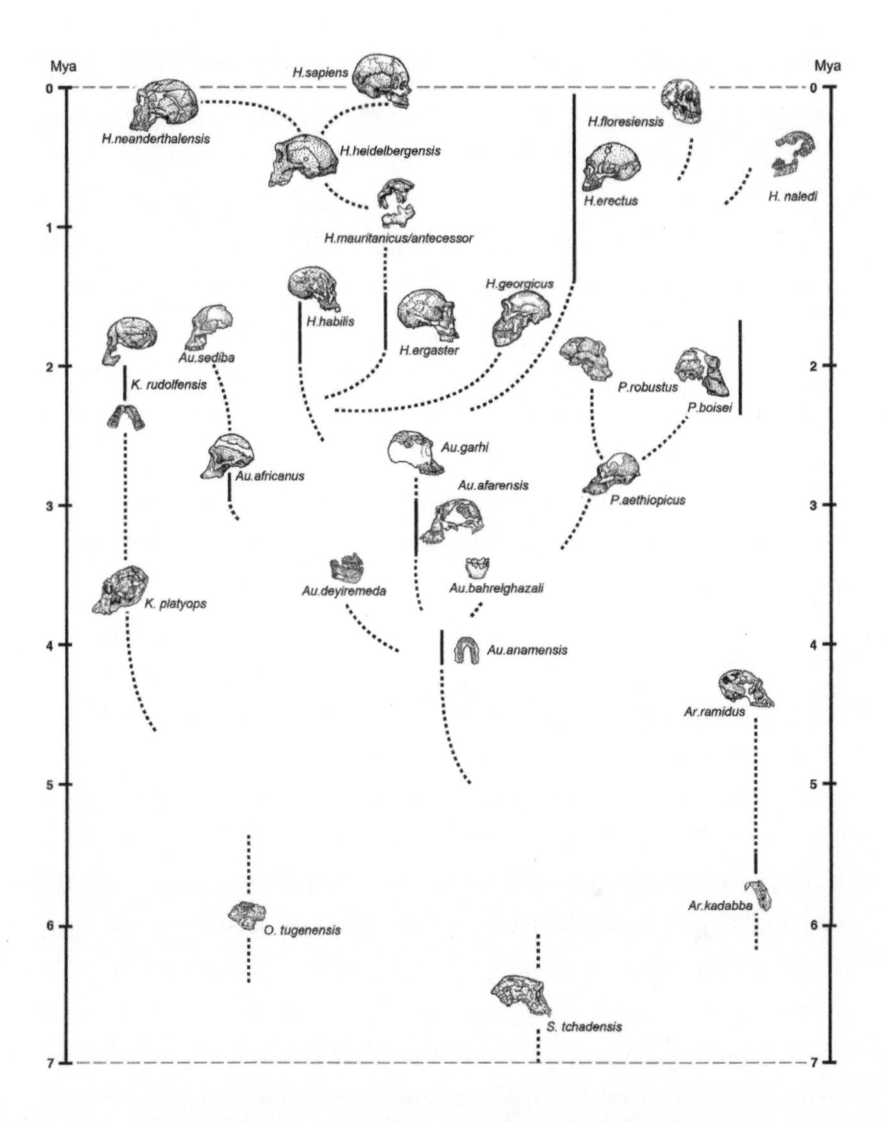

Very tentative family tree of the hominids showing substantial diversity, frequent branchings, and the typical coexistence of multiple species. *Drawn by Kayla Younkin.*

With fossil representatives known from the period between about 3.8 and 3 million years ago, *Australopithecus afarensis* appears to have occupied a mosaic of habitats that ranged from dense woodland to bushland to tree savanna, and that were nearly always close to sources of water. This matches well with what we know of the skeletal anatomy of Lucy and her like, who unquestionably walked upright when on the ground, as testified by their broad, shallow pelvis—the very antithesis of its tall, narrow equivalents in the apes—and their leg and foot anatomy. The foot of *A. afarensis* is particularly interesting in this regard: it is relatively long, but the toes are all in-line, just as ours are, and both longitudinal and transverse arches were present. However, the anatomy above the waist tells a somewhat different story; long arms, an upwardly tapering thorax, the narrow shoulders, and the longish, slightly curved hands of *A. afarensis* all speak of great facility in the trees (though it has been suggested that Lucy herself died falling out of one). All in all, it seems that in coming to the ground the australopiths had not cut themselves off from their ancestral habitat: instead, they retained many apelike climbing abilities while at the same time favoring upright locomotion when away from the trees. Most authorities now subscribe to the view that the australopiths lived an existence distributed between the arboreal and terrestrial milieus: feeding, sleeping, and taking shelter in the trees, even as they traveled on the ground and exploited the resources that the earth-bound environment had to offer.

This interpretation makes a lot of sense. For, while the australopiths were about as big and invulnerable (in the trees) as any arboreal mammal usually gets, coming to the ground exposed them to a dangerous environment that teemed with large and ferocious predators. This was hardly a comfortable place for a slow-moving and still relatively small-bodied mammal that lacked sharp, slashing canine teeth to defend itself with. So how did the australopiths cope? Possibly, they sought safety in numbers. One remarkable feature of *Australopithecus afarensis* is the apparently very large size difference between males and females, the former standing around 150 cm (4 feet, 11 inches) tall, while the latter come in at an estimated 105 cm (three feet, six inches). This corresponds to an estimated 40 kilos (90 pounds) versus 30 kilos (66 pounds) in body weight. In the admittedly small sample of living African ape species, extreme sexual dimorphism of this kind (though interestingly not seen in the canine teeth of *A. afarensis*, as it is in gorillas) is associated with small social groups, in which a dominant male monopolizes the reproductive females; lower dimorphism is seen in species that live in larger groups in which adult males compete for access to females. Almost certainly, though, this does not imply that *A. afarensis* lived in gorilla-like troops. Instead, it has been cogently argued that when we try to infer the social organization of the early hominids, we should be making comparisons not with their closest living relatives but with their closest living ecological equivalents, namely the baboons and macaques. Exposed to similar dangers (but faster, and with fearsome canines), these quadrupedal monkeys live in very large groups in which the crucial reproductive core is protected on its exposed periphery by the presence of large numbers of less reproductively valuable individuals, such as subadults or nondominant males.

If the australopiths did anything similar as they accommodated to the hazardous terrestrial milieu, they would have departed substantially from the ancestral ape pattern, both demographically and in the ways in which their societies were organized. And while that is largely speculation, we have much stronger evidence that the australopiths diverged markedly from the apes in terms of their economic strategies. When modern chimpanzees emerge from the forest into savanna environments, as they sometimes do, they continue to forage for basically the same resources that sustain them in the trees: mainly fruit and leaves, supplemented by the occasional captured small vertebrate. In contrast, under similar environmental circumstances australopiths evidently began to exploit a new range of resources not found in the forests—tubers, fleshy roots, rhizomes—when they found it necessary to do so, although the way their teeth wore down suggests that they might still have preferred fleshy fruit when available. This dietary innovation seems to be confirmed by analyses of stable chemical isotopes that leave a chemical signature in the teeth that differs according to whether the plants consumed are of the bushy or grassy sort. They suggest that *Australopithecus afarensis* differed from its presumed antecedent *A. anamensis* in showing a chemical signature of a kind associated with resources predominantly available in more open environments.

It is notable that this chemical signal may also be acquired not from the vegetation itself, but from eating grazing animals that feed on grasses, sedges, and so forth. And this leaves open the possibility that *A. afarensis* was an opportunistic feeder that may occasionally have eaten a small mammal or other vertebrate. This would hardly be surprising, since today's chimpanzees are well known for their propensity to cooperatively hunt colobus monkeys, small bush pigs, and so forth, and to share their meat. To apes, the importance of such behaviors is probably more social than nutritive, reinforcing reciprocal relationships; but it is entirely plausible that early australopiths may have at least occasionally consumed meat in the context of a highly generalist dietary strategy—although their broad pelvises may at the same time suggest that they possessed capacious and generally apelike digestive systems.

What behavioral innovations of this kind may have meant in cognitive terms is hard to say. The australopiths have been described as "bipedal apes" because, although they walked upright, the proportions of their skulls are very apelike, with a large, projecting face dominating the small braincase behind it (quite the opposite of *Homo sapiens*). But their novel fashion of relating to the environment around them nonetheless suggests that they were beginning to process information about the world in a new and different way. With a mean of somewhere around 450 ml, australopith brains were on average close to 25 percent greater in volume relative to body size than those of chimpanzees and the very early hominids; and in some very general way this must reflect an increase in something one would have to call "intelligence," although just what that might have meant in terms of how individual australopiths apprehended the world around them is impossible to specify. The same caveat also applies to the first early hominids who spontaneously began to make stone tools; but that new activity was nonetheless even more emblematic of an exit from the ape cognitive realm.

THE EARLIEST STONE TOOLMAKERS

It is only with the invention of the first durable (hence preservable) technology—the stone tool—that we begin to have material evidence of a revolutionary change in hominid lifeways. By any standards, manufacturing stone tools was a radical innovation; but it seems nonetheless to have been introduced by creatures that were anatomically and otherwise archaic. The australopiths had certainly departed from the ancestral ape condition in having adopted both terrestrial bipedality and a more generalist ecological strategy; but they could hardly be said to have been actively transitional between the apes and ourselves. To the contrary: their apparent general uniformity in body form over an extended period suggests that they had adopted a unique and apparently stable lifestyle, one that endured for several million years even as its practitioners experienced significant environmental changes on both shorter and longer scales.

Yet, just as terrestrial bipedality emerged among apes, it was among the australopiths that the most foundational human behavioral uniqueness of them all appeared: the first significant step along the long road that eventually led to the technology-driven modern human domination of the world. That might seem a little odd, for wouldn't you expect an innovation as radical as the deliberate manufacture of stone tools to be introduced by a new kind of hominid, maybe one with a bigger brain? Well, actually no; for this is, in fact, an excellent example of a pattern that has held throughout the human evolutionary story, in which innovations in the biological and technological realms have been consistently out of phase.

This disconnect actually makes good sense, because any behavioral novelty has ultimately to be introduced by an individual who necessarily belongs to a preexisting species. What is more, any new behavior can be adopted only if the biology that permits it is already there, underlining the importance in human evolution of exaptation, the routine (and perhaps even obligatory) evolutionary process whereby new inherited features come into existence at random, entirely unrelated to any future new function to which they might be co-opted—even if they necessarily represent variations on what is already there. The classic example of exaptation is feathers, which kept ancestral birds warm long before they were recruited for flight. But the biological histories of organisms in general, and of humankind in particular, are packed with others.

Following the 3.4-million-year mark, a point at which australopiths were the only hominids around, we find a couple of rather anomalous occurrences in the nascent archeological record. From a place called Dikika, in Ethiopia, paleontologists recovered some pieces of mammal bone bearing long incisions that their finders believed had been made by the cutting edges of sharp stone tools. And from Lomekwi in Kenya, only slightly later in time, some curiously large shards of battered stone are claimed to be the work of early hominid knappers. Still, both of these putative technological manifestations are outliers, and so for the moment both lack any larger context in which they can be placed.

The earliest unambiguous occurrences of stone tools, again in Ethiopia and Kenya, come from rather later, around 2.6 million years ago—following which stone artifacts become a common feature of eastern and southern African sites. Given the name "Oldowan" from the site in Tanzania at which they were first identified, these earliest stone tools are pretty unimpressive to look at, consisting of fist-sized stony cores and the small sharp flakes that early hominids knocked off them using another stone as a hammer. But that sharp cutting edge made all the difference! For the first time, hominids without the specialized puncturing and slicing teeth of carnivores could efficiently and rapidly butcher a mammal carcass, always a major attraction for dangerous predators and scavengers of various kinds. It's been suggested, in fact, that a major advantage of the early use of cutting tools would have been the ability to remove prized carcass parts so that they could be taken to less hazardous places for consumption.

This is not to say that at 2.6 million years ago hominids were already becoming specialized meat-eaters; and indeed, even today, as the world's top predator, we human beings are still conspicuously generalist in our dietary preferences. The early toolmakers probably still depended on a wide range of plant resources for sustenance; but by inventing butchery they had opened a door, and there was no going back. The hominids who ushered in this revolution were almost certainly australopiths, because we have no convincing fossil evidence that any more advanced kind of human precursor was around at that point in time. What is more, the only putative association we have at this early stage between tool-using activity and a specific kind of hominid involves a 2.5-million-year-old form from Ethiopia called *Australopithecus garhi* and cut-marked bones.

The use and even the manufacture of tools are not in themselves uncommon things. New Caledonian crows, for instance, are famous for using long thorns held in their beaks as aids in foraging, and sea otters use stones to smash open mollusk shells. Among the primates, capuchin monkeys, long-tailed macaques, and chimpanzees have all been observed using rocks as hammers and anvils for cracking open hard nuts; and now gorillas as well as chimpanzees have been observed using stripped-down plant stems to "fish" ants out of their nests. Chimpanzees in West Africa have even created their own "archaeological record" by virtue of the discovery of stone-pounding tools at a site dated to 4,300 years ago: something that dramatically confirms the long-term cultural transmission—the passing down by learning through the generations—of their nut-cracking behaviors.

Nonetheless, the spontaneous invention of techniques to manufacture sharp-edged stone tools is something that is pretty plainly beyond the cognitive capacity of any living ape. Intensive experiments involving Kanzi, a bonobo who had performed impressively in "ape language" experiments, showed that he quite readily got the idea of smashing stones to obtain rewards; but he never grasped how to strike one rock with another at precisely the angle and force necessary to achieve a sharp flake. This limitation appeared to have its origins in what was going on in Kanzi's head, rather than in the undisputable fact that his long, slender, grasping, short-thumbed hands did not have the powers of precise manipulation possessed by the broader and shorter hands of early hominids. What is more, it is pretty certain that Kanzi and other living apes do not, even in a rudimentary manner, understand that not every kind of stone is equally suited for knapping, while in contrast Oldowan hominids routinely carried suitable cobbles around with them for miles, evidently in anticipation of needing them as raw materials in places where the right stones might not be available.

HOW EARLY HOMINIDS "THOUGHT"

The first hominid stone tool manufacturers had an unquestioned exaptive advantage in possessing the ability to adopt stone toolmaking. And putting all the evidence together makes it clear that the ability to make stone tools is not simply a manual skill, but that early toolmakers had entered entirely novel cognitive territory. Quite evidently, these early hominids did not see the world through ape eyes. Unfortunately, though, that's about as much as we can say on the subject. The reasons for this are varied. Very importantly, while we can observe what apes do in various situations, and even make predictions about how they may react under particular circumstances, we can never know precisely how an ape is subjectively experiencing the world. We can make broad-brush judgments about ape emotional states—whether an individual is happy or depressed, for instance—because we experience those things using roughly the same brain circuits that apes possess. But it is very clear that apes put together what they perceive about the world very differently from us, and that this difference makes it tough for a human being to judge exactly what an individual ape is mentally undergoing at any specific moment in time.

Partly, this is because mental states are largely interiorized; but to an even greater extent it is because we ourselves are literally the prisoners of our own peculiar cognitive style, seeing everything through the filter of the modern human intellect. This makes it hugely tempting for us to see early hominids—our relatives, after all—merely as simpler versions of ourselves; but to give in to that temptation would be gravely misleading. We can't hope to understand or even broadly characterize our predecessors' cognitive processes by simply knocking, say, fifteen IQ points off the average hominid intellect for every million years we go back into the past. There are many reasons for this, but the main one is that, compared to all our mammal, primate, and hominid relatives, there is something *qualitatively* unusual about the way in which we *Homo sapiens* process information. We don't just do the same thing better, or quicker. We do something completely different. And while this difference confers some very impressive abilities on us, it comes with severe limitations as well. One of them is that, quite simply, we are incapable of precisely imagining cognitive states other than our own: cognitive states that lie on the other side of the barrier that, at some point, our ancestors crossed. And, perhaps paradoxically, the attempt becomes more difficult the more closely we are related to the subjects of our speculation. For it is literally impossible for us to envisage a form of consciousness that is almost like ours—but not quite.

Still, whatever the details of the way in which they apprehended the world, it is safe to conclude that those early hominid toolmakers had made a significant cognitive leap, relative both to the living apes and to their own ancestors. They were processing information about their environments in an unprecedentedly complex manner and, to that extent, they were closer to us than any other organism had ever been. But they were not *us*, nor even a primitive version of us. For while they were responding to the world around them in an unusually nuanced way, there is no indication whatever that they had begun to reason in anything even resembling the modern human manner.

It is a good bet that members of every species alive today, *Homo sapiens* apart, experience the world around them as an existential whole. And they react to that whole intuitively—in more or less sophisticated ways—perhaps somewhat as a modern human might to an Impressionist painting. This was evidently also true for the australopiths. In contrast—and apparently uniquely in the history of life—we modern humans deconstruct both our interior worlds and the physical and social environments around us into a vocabulary of discrete mental symbols. When those symbols have become lodged in our minds, we can combine and recombine them in new ways, according to rules, to envision alternative versions of the worlds we experience, or even to conjure up entirely imaginary ones. One hugely resonating result of this odd capacity is that we human beings live at least as much of the time in the worlds we reconstruct in our heads as we do in the world that nature directly presents to us. This makes us truly unusual, presenting us, among many other conundrums, with the familiar problem of explaining the unique.

Now, you will probably find this brief characterization of what it is that makes our modern human cognition unprecedented less than entirely satisfactory. And, quite frankly, so do we; for in the end, the notion of symbolic cognition is no more than a useful shorthand for something that we do not yet fully understand about the way in which we humans process information—and that we will never properly comprehend until we can explain exactly how a mass of electrochemical discharges in the physical brain becomes resolved into what each of us experiences as his or her individual consciousness. The wait promises to be a long one; but if we are to try to reconstruct the evolutionary context within which our modern consciousness emerged, it is vital to have a placeholder of some kind—and as we will see, it turns out that the notion of symbolic consciousness has several major advantages in this respect. In any event, for all the many striking physical peculiarities of *Homo sapiens*, it is clearly our cognitive capacities that most sharply set us off from even our closest relatives in nature. And for this reason, much of the rest of this chapter will be devoted to looking in the fossil and archaeological records for indicators that those capacities had been acquired.

THE EARLIEST *HOMO*

Paleoanthropologists debate endlessly what the boundaries of our own genus *Homo* are. Historically, this cluster of related species has grown by the steady accretion of new members, until nowadays it has become a bit overstuffed for our taste: for extreme enthusiasts, it includes forms with large chewing teeth (and otherwise basically unknown) that date back almost three million years. Yes, very probably our lineage is ultimately descended from an as yet undiscovered form that we would have to classify as australopith if we knew it. But if we rein ourselves in a little, and sensibly restrict our genus to fossil hominids that are both reasonably closely related to *Homo sapiens*, and that share significant aspects of cranial anatomy and body form with us, we find that the fossil record offers us the earliest plausible contender for membership in *Homo* at a little under two million years ago.

Still an exclusively African form, that contender is the species *Homo ergaster*, best known in the guise of the famous Nariokotome Boy from northern Kenya. The Boy is the remarkably complete and well-preserved skeleton of an adolescent who died about 1.6 million years ago at the tender age of eight, but who had been at the stage of development of a modern thirteen-year-old (we modern humans are very unusual in our slow rate of maturation, a trait that is probably related to the very long period we spend learning about the world). Although recent studies have shown that this individual was a little less advanced in his body structure than had initially been believed, the Boy showed significant modifications in the modern human direction when compared to the australopiths. Most importantly, at around 160 cm (five feet, three inches) he was already a lot taller than any australopith, and he was quite slenderly built, with long legs that would have suited him well for striding and trotting across the open savanna. At last, in the adolescent from Nariokotome we have evidence of a hominid that was capable of living permanently out in the open, away from the shelter of the trees—but that was, commensurately, a less agile climber. To revert to a bit of jargon, the Boy tells us that *Homo ergaster* was very clearly an *obligate* biped, rather than simply a *habitual* one.

This commitment to the expanding savanna biome involved a change of environment as radical as the original descent from the trees reflected in the shift from facultative to habitual bipedality. For one thing, away from the shade of the woodlands hominids were faced with an entirely new problem: shedding the heat load imposed by the merciless tropical sun. This issue was a major one, because the internal organs of the body—and most especially the brain—are exquisitely temperature-sensitive. If the brain overheats, even for a short while, the consequences can be disastrous. And for a hominid lacking any specialized mechanisms for cooling the brain, the only solution to the problem lay in keeping the entire body cool. In part, those early members of *Homo* were inadvertently (or exaptively) helped in this goal by their vertical posture. This exposed the smallest possible heat-absorbing body area to the rays of the sun directly overhead, while at the same time maximizing the vertical and thus heat-losing surfaces of the body that were raised above the hot ground and exposed to cooling breezes.

Diorama at the American Museum of Natural History, showing two *Homo ergaster* in Kenya's Turkana Basin at around 1.8 million years ago. Note that despite their substantially modern body structure they are using tools of the kind already made for almost a million years by their australopith predecessors. *Courtesy of AMNH.*

But by itself this postural adjustment wouldn't have been enough. The major mechanism that the human body uses to cool itself is the evaporation of sweat. And that evaporation would have been impeded by the hairy coat that early hominids inherited from their ape ancestors. It is unknown how much of that thick ancestral coat the australopiths retained in their woodland setting, although it seems likely that some was. But for the early members of the genus *Homo*, out there on the savanna it would have been imperative to get rid of what hair remained (except for a protective blanket over the crown of the head). This loss would, of course, have directly exposed their delicate skin to the damaging ultraviolet radiation of the sun, dictating that protective dark pigment be expressed in its superficial layers.

All in all, the new environment demanded huge accommodations—behavioral as well as physiological—from the hominids that moved in to occupy it. And it was not simply a new environment; it was a new ecological niche. A niche that on the one hand teemed with dangerous predators, and that on the other demanded at least a degree of predatory behavior. For it was at this point in human evolution that the hominid brain began seriously to expand in volume. And brain is a pretty voracious tissue, in our case accounting for some 20 to 25 percent of all the energy we consume despite constituting only about two percent of our body mass. This means not only that you don't need any more brain than the absolute minimum you can get away with, but that, if you are going to add more brain, you will have to have a bigger or better diet to supply all the extra energy it will demand. The first hominids had ventured away successfully from the ancestral forests with brains only very modestly larger than those of the apes; but the new open savanna environment was going to make vastly greater demands on the brains of their successors, and on the behaviors those brains made possible.

Ape diets are of relatively low quality. That is, they do not yield a lot of energy per unit of food consumed; and it seems likely that ape brains—already large by both primate and general mammalian standards—are about as big as their fruit-and-leaf diets will support. The australopiths' brains were marginally larger than the apes'; but one of the many striking things about the Nariokotome Boy is that his brain was almost double the size of the average australopith's, even though his body weight, at an estimated 48 kg (106 pounds), was within the range of chimpanzees. The actual volume of his braincase was some 880 ml; had he survived to adulthood, it would have exceeded 900 ml. Maintaining a brain of this size would have necessitated a diet of radically improved quality, and the only evident source of such a diet was the mammal carcasses that had long been at least occasionally scavenged by the australopiths. But scavenging, even the power-scavenging—the chasing of rivals off carcasses—that many believe the australopiths must have indulged in, would in all likelihood not have provided an adequate source of the animal fats and proteins needed to support a brain as large as the Boy's. Some active hunting would almost certainly have been required as well.

But wait; there's more. The capacious intestines of apes (and almost certainly of australopiths, too) are well suited for the processing of large amounts of relatively nutrient-poor food such as leaves and unripe fruits. But when a chimpanzee has killed and eaten a monkey, it will excrete a large amount of poorly digested flesh because its gut simply cannot efficiently extract the energy potentially available. And while the Nariokotome Boy very probably had a gut that was greatly reduced compared to an australopith's, he certainly did not have the specialized digestive apparatus of a dedicated carnivore. It is for this reason that the primatologist Richard Wrangham has argued, quite persuasively, that once *Homo ergaster* was out on the savanna, its members would necessarily have depended on the cooking of both animal and vegetable foodstuffs if they were to extract the level of nutrition they and their brains needed.

The cooking of food using controlled fire clearly would have had numerous advantages. It would have made both animal and plant foodstuffs easier to chew, especially for a creature with a significantly reduced face and dentition; it would have made much more of the nutrition contained in the food available to the gut; and it would have detoxified flesh that rapidly rots and becomes poisonous in the tropical sun. What's more, Wrangham emphasizes the dangers that, especially at night, would have lurked out there on the savanna for a creature without the natural defenses of a specialized predator. Controlling fire, he points out, would have been an excellent mechanism for discouraging large carnivores from taking too close an interest.

However you may choose to look at it, the domestication of fire was clearly a momentous development in the human behavioral repertoire, and it has clear implications not just for the physiology and the individual cognition of the hominids who first practiced it, but for the organization of their societies. Fire provides a unique focus for the social group and for its activities; and unlike the australopiths, as a secondary predator *Homo ergaster* most likely lived in groups of limited size that could be protected by a single hearth. Put everything together, and a picture begins to emerge of a highly social but relatively sparsely distributed creature that was significantly in the mold of later hominids. Still, as tempting as it might be, it is premature to let this picture become too specific.

For a start, while *Homo ergaster* probably committed itself to the savanna environment not long after two million years ago, we have no actual evidence for the control of fire in hearths until almost a million years later. South Africa's Wonderwerk Cave has yielded a succession of ancient hearths that evidently housed controlled fires about a million years ago; but evidence of this kind is sporadic at best until well under half a million years ago, after which fire finally became a familiar feature at early hominid occupation sites. Of course, it is always possible that the dearth of early evidence for fire control is due to issues of preservation over the eons. But until this can be determined, or until the record improves, Wrangham's argument, attractive as it is, will necessarily remain a circumstantial one.

Still, little as we can confidently say about the day-to-day lives of those *Homo ergaster* out there on the savannas, we do know that the radical habitat shift imposed a whole lot of new exigencies on these hominids. And they must have responded in an appropriate manner. By this point, hominids had been using simple stone tool kits for a long time, and one response you might expect would be a change in stone-working technology. But interestingly, no such thing seems to have occurred. The very first *Homo ergaster* we know apparently continued to make and use the same sharp flakes and pounding implements that their more physically archaic predecessors had already been using for over half a million years. With one single outlier at 1.78 million years ago, it was not until after the Nariokotome Boy had lived and died around 1.6 million years ago that a new stone-working technology became widely adopted. This was a revolutionary new conceptual development in stone toolmaking that is known as the "Acheulean handaxe."

Prior to the advent of the handaxe, it had not mattered what a stonecutting tool looked like. The only requirement was that it have a sharp edge, and most of the resulting sharp flakes were only a few inches long. What's more, such implements were typically made on an as-needed basis, knapped when and where they were used. The much larger handaxe changed all this. It was the first kind of stone tool that was worked on both sides to conform to a consistent (usually teardrop) shape that the toolmaker must have had in mind before knapping began. And eventually, handaxes were often manufactured in vast quantities at central "workshops." Sometimes they were even made with an element of playfulness, as when the toolmakers apparently competed to make the largest possible implements, regardless of how heavy or impractical they might have been.

Over time, handaxes became slimmer and typically more elegant. But this "Swiss Army Knife of the Paleolithic" persisted, unchanged in basic concept, for well over a million years. And, once again, it was an innovation introduced during the tenure of an established hominid species. The handaxe encapsulated an entirely new way of looking at the potential lurking within a piece of stone; but while one might well imagine that this revolutionary insight might have necessitated an entirely new cognitive mode, the first handaxe-maker almost certainly looked and behaved pretty much like any other *Homo ergaster.*

Sadly, the record does not currently allow us to say much more than this about *Homo ergaster* as a cognitive being. We know that this hominid's way of life, its social organization, its economic strategies, and almost certainly its way of apprehending the world, must have been different from those of its australopith ancestors. But we have no idea just how different they were, or even if the changes involved were acquired short-term or sporadically accumulated over time. We just know that, with *Homo ergaster*, a new kind of hominid had emerged: one that ultimately gave rise to one of the most interesting evolutionary experiments of all time: the diversification of the genus *Homo* and its initial spread throughout the Old World. The archaeological record associated with this pivotal species suggests that it had reached a level of technological and presumably cognitive sophistication greater than anything that had preceded it. But, significantly, nothing in that record suggests that *Homo ergaster* was even anticipating the curious style of information processing that is the most striking hallmark of humanity today.

HOMO HEIDELBERGENSIS

The eastern African fossil record that gave us *Homo ergaster* fades a bit after around 1.4 million years ago—which is also when the last (robust) australopith disappears. Scattered fossils indicate that a fair amount of diversification was happening in early African *Homo* in this general period, but what exactly was going on remains hazy. Better documented is the fact that hominids had managed to leave Africa for the first time by about 1.8 million years ago, the date of a remarkable assemblage of hominid fossils from the Republic of Georgia, in the southern Caucasus. What is more, early dates indicate that hominids had penetrated all the way to eastern Asia not long thereafter, as testified by fossils from both China and the Indonesian island of Java. By around 2 million years ago, then, the hominid story was enlarging to embrace much of the Old World, including the invasion of the environmentally novel cool temperate zone. Clearly, those peripatetic hominids were highly adaptable and behaviorally flexible.

Still, the first hominid species to be found not only in tropical Africa but across the Eurasian continent did not emerge until significantly later, also from an ultimately African ancestry. This first cosmopolitan species was

Homo heidelbergensis. It was named in 1908 for the German city near which the first specimen was found, but representatives are now known also from sites in southern and eastern Africa, as well as from elsewhere in northern and southern Europe and as far afield as China.

Homo heidelbergensis seems to have fairly closely resembled *H. sapiens* in its limb proportions, though the individual bones of its skeleton were more robustly built than ours typically are. At an average volume of around 1,260 ml its brain was a lot larger than that of *H. ergaster*, though it was still somewhat smaller than typical of later hominids such as *H. neanderthalensis* and early *H. sapiens*. That energy-hungry brain was housed in a long, low, and usually thick-boned cranial vault set behind a robustly constructed face topped by tall, swelling brow ridges, and the jaws of *H. heidelbergensis* housed chewing teeth that were significantly smaller than those of its predecessor. This cranially distinctive hominid is now known from sites about six hundred thousand years old in both Africa and Europe, though many of its (typically poorly dated) representatives may be significantly younger than this.

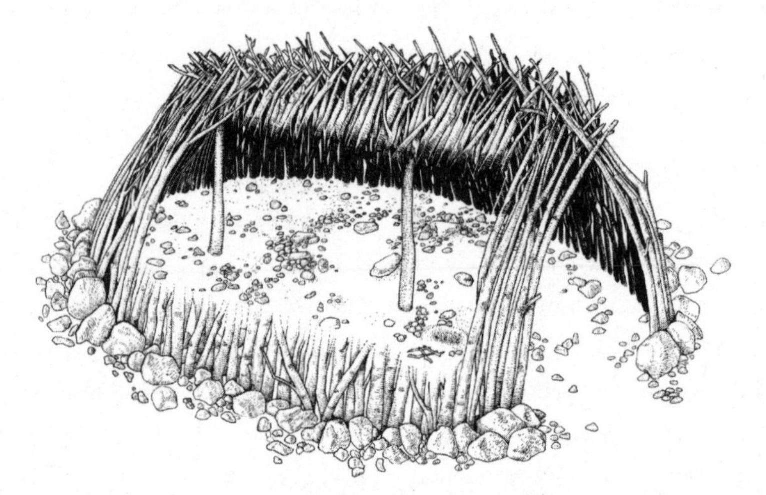

Reconstruction of a hut built around 380,000 years ago at the French Mediterranean site of Terra Amata. Saplings were embedded in the ground around a large oval periphery and brought together above. Visible just within the gap believed to represent the entrance is a shallow pit, within which a controlled fire had burned. Weatherproofing would have been with dense brush or possibly hides. *Drawn by Diana Salles.*

In a very general sense, *Homo heidelbergensis* is often considered to represent the ancestral stock from which the European *H. neanderthalensis* and the tropical African *H. sapiens* lineages diverged, and over its tenure (generally reckoned to have been between about seven hundred and two hundred thousand years ago) several very significant technological developments are documented, mainly in Europe. One of the most interesting of those developments is the deliberate construction of early artificial shelters, best known at a site on the French Mediterranean coast called Terra Amata. On an ancient beach, some 380,000 years ago, a group of hunter-foragers built a camp consisting of several large huts. Each consisted of an oval arrangement of saplings that were sunk in the ground, reinforced with a ring of stones, and then brought together at the top. Whether or not this scaffold was wind- and waterproofed with hides is unknown for sure; but just inside the gap in the stones that presumably represented the best-preserved hut's entrance, excavators found a scooped-out hearth containing burned animal bones and lined with fire-blackened pebbles. Very little that is comparable is known from before this time, but following Terra Amata times rudimentary structures became increasingly frequent features at archaeological sites. Many earlier campsites are known, of course, often occupying naturally sheltered localities such as cave entrances and rock overhangs; but the Terra Amata huts are the best early evidence we have for constructed dwellings with their inevitable overtones of "home"—even for the itinerant hunter-gatherers of the age.

During the tenure of *Homo heidelbergensis* we also find evidence of the first compound tools—stone points hafted into wooden handles. The best evidence comes from the German site of Schoeningen, some three hundred thousand years old, at which soft materials that would normally have rotted away have been preserved in an oxygen-poor boggy setting. As anyone who has ever tried to use a saw blade without a handle knows, the hafting of implements vastly increases their effectiveness, and the advent of the compound tool represented a major technological leap. The finding in 2013, at South Africa's half-million-year-old Kathu Pan 1 site, of stone points bearing indirect suggestions of original hafting suggests that this technique had been employed long before Schoeningen (but not before *Homo heidelbergensis*) times. Schoeningen, however, provides us with the first actual material evidence. Schoeningen has also produced the world's earliest known wooden throwing spears, carefully shaped like javelins with their weight concentrated at the front. How effective these spears' fine wooden points would have been at penetrating the tough hides of grazing animals has been questioned; but there's no doubt that whoever made them had a fine intuitive understanding of missile physics.

It was also during the time of *Homo heidelbergensis*, at some point after around four hundred thousand years ago, that the first conceptual innovation in stone toolmaking since the handaxe was made. This was the introduction of the "prepared-core" tool, in making which a suitable piece of stone was carefully shaped with blows to both sides, until a final sharp impact would detach a flake that was either a finished tool in itself or that provided a blank that could be shaped as desired. The big advantage of this technique was to provide a continuous sharp cutting edge all around the periphery of the implement, and its advent suggests that a further cognitive leap had been taken by ancient stone toolmakers who added yet another level of abstraction to the manufacturing process.

With an impressively sophisticated range of technologies at their disposal, it is hardly surprising that members of *Homo heidelbergensis* appear to have been formidable hunters of large animals. At the half-million-year-old site of Boxgrove, England, bone fragments presumed to represent *H. heidelbergensis* were found along with the cut-marked (thus butchered) bones of horses and woolly rhinoceroses, while Schoeningen seems to have been a virtual abattoir of horses. You might have expected to find this degree of hunting sophistication in a hominid of quite impressive brain size such as *H. heidelbergensis*, but what is perhaps more surprising is that this species left us virtually nothing that we might reasonably associate with even the rudiments of the modern symbolic way of processing information.

A conceivable exception to this is a lumpy volcanic pebble from the site of Berekhat Ram, in the Golan Heights region of the Middle East, that is dated to about 230,000 years ago. This small piece of stone is vaguely shaped like a female torso, and it bears three shallow grooves that may have been deliberately made to enhance that resemblance. It is perhaps matched by an intriguingly shaped and possibly even more ancient lump of quartzite found at the Moroccan site of Tan-Tan and that might (or might not) have been intentionally "improved" to resemble a female figure. But while both these objects are suggestive, each is contested as a work of deliberate symbolic intent; and whatever the case, they are floating points without any wider symbolic context into which they might be fitted. Interestingly, the same thing can be said for a half-million-year-old mollusk shell, putatively associated with the eastern Asian species *Homo erectus*, that bears a curious sequence of zigzag incisions. Whoever made those incisions certainly possessed the manual skills to perform the engraving, but the obvious issue of intentionality is something that must remain in abeyance until the record improves.

Putting together all we know about *Homo heidelbergensis* and its putatively associated record (and we wish we knew a lot more) thus yields a very vague portrait of a hardy, ingenious, and resourceful hominid that flourished in a wide variety of habitats, and that responded to its environment in sophisticated ways. What it does *not* yield, though, is a portrait of a species that we can ever hope to comprehend as a primitive version of ourselves. *Homo heidelbergensis* almost certainly did not see the world through modern eyes, or even show the rudiments of doing so in any substantial way. An amazing success in its own time, this hominid evidently had its own very successful fashion of doing business. But the main lesson it seems to teach us is that there is a variety of ways of being a clever hominid—and that, by extension, our own specific way of being clever is only one of many.

THE NEANDERTHALS

Without question, the best-known of all extinct hominid species is *Homo neanderthalensis*. The Neanderthals emerged in Europe or western Asia some two hundred thousand years ago, from local antecedents who had already been occupying this region for several hundred millennia. They became extinct soon after the first *Homo sapiens* (often known as Cro-Magnons for the place in western France where their skeletons were first uncovered) began arriving in their homeland a little over forty thousand years ago, from a point of origin somewhere in Africa. In the interim, the Neanderthals left us an incomparable record of a highly evolved and sophisticated hominid species that, over an extended period, contrived to successfully weather some extreme environmental fluctuations. What permitted this success was a complex culture and technology that compensated for the severe circumstances in which Neanderthals often lived—although at times during the Ice Ages that afflicted Europe and western Asia during their tenure, they did find themselves having to abandon wide swaths of territory.

Clothing would have been paramount among the Neanderthals' cultural accommodations to cold environments, for it has been estimated that, without added insulation, during the coldest periods adult Neanderthals would have required more than 50 kg (110 pounds) of subcutaneous fat to keep their internal organs at working temperature. Hardly an efficient option for an itinerant hunter-gatherer! No ancient clothing—which presumably took the form of perishable animal hides—has survived; but scientists using DNA have ingeniously estimated that the two species of lice that inhabit the human head and human clothing speciated between 80,000 and 170,000 years ago, suggesting on this basis that early tropical *Homo sapiens* may already have been experimenting with clothing before leaving Africa to take over the rest of the world. This date range also implies that clothing would have been an independent Neanderthal—or even earlier—introduction that we have no means of calibrating.

What we do know is that the modern humans who displaced the Neanderthals were the first hominids to wear tailored clothing, as attested by their invention of sharp, tiny-eyed bone needles at some time before about twenty-seven thousand years ago. How Neanderthal (and earlier *sapiens*) clothing was held together remains a mystery, though sinew thongs passed through holes punched in animal skins are an obvious possibility.

The Neanderthals had brains as large as those of the early *Homo sapiens* who replaced them (and larger than ours nowadays). But those brains were enclosed in skulls of distinctly different shape. In contrast to our globular braincases, those of the Neanderthals were long and low, ending in a prominence at the rear; and, unlike our small, retracted faces, their somewhat larger ones projected forward and were capped by arching brow ridges above each eye. *Homo neanderthalensis* was quite variable in stature, some individuals achieving 183 cm (6 feet); but the bones of Neanderthal skeletons are typically more robust than ours, and their conical rib cages flared below to match broad pelvises. On the landscape, the two species would have presented subtle but distinctive differences in aspect. But while Neanderthal bones are characteristic enough to make them fairly easy to recognize as fossils, in general their possessors seem to have been pretty average hominids for their time. For example, they clearly shared their general skeletal robustness with species such as *H. heidelbergensis*. It is the lightly built and slender *H. sapiens* that departs most strikingly from the more robust ancestral condition.

Neanderthals were masters of the prepared-core stoneworking technique. Many of their "Mousterian" tools were exquisitely made, and they were modified by a process of reduction into a whole variety of purpose-specific implements: scrapers, knives, piercing tools, and so forth. These were often mounted into handles. Perhaps oddly, though they carefully curated valuable pieces of stone that kept good edges, Neanderthals do not seem to have made much use of softer materials such as bone or antler (though one curious bone piece, evidently used as a "retouch" implement for finishing stone tools, was made from a piece of hominid skull). Neanderthals routinely burned fires in hearths at their campsites, and at least occasionally they rigged up shelters, although they tended to discard their garbage rather indiscriminately, instead of disposing of it outside their living spaces as the Cro-Magnons habitually did. Perhaps even more suggestively, it is pretty clear that at least occasionally, and very simply, the Neanderthals buried their dead: a habit that accounts for the remarkably good fossil record we have for them.

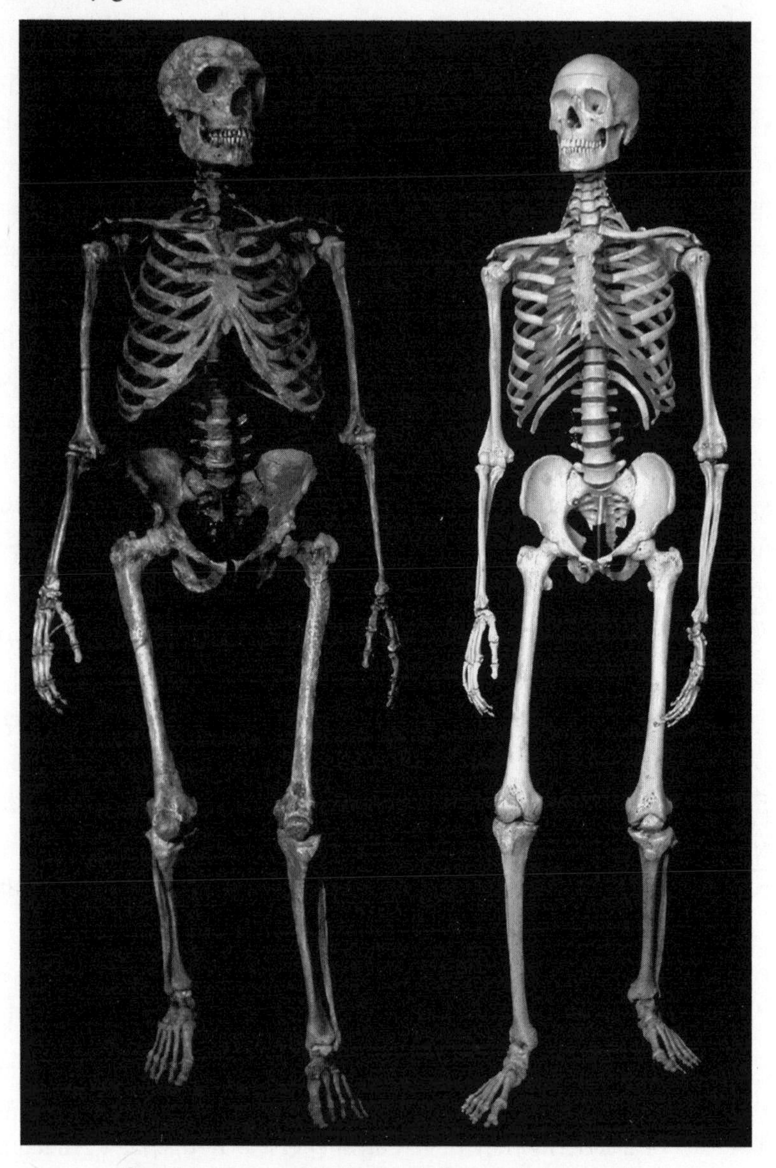

Comparison of a reconstructed composite Neanderthal skeleton with a *Homo sapiens* of similar stature. Note the extensive differences, especially in the cranium and the thoracic and pelvic regions. *Photograph by Ken Mowbray.*

To judge from the extent of the typical Neanderthal living site, social group sizes were quite small (certainly compared to those of the Cro-Magnons, who seem sometimes to have come together into groupings of several hundred). At one site in Spain, an entire social group of Neanderthals is believed to have been massacred and subsequently butchered, presumably by other Neanderthals, among whom cannibalism appears not to have been an infrequent practice. The unfortunate group consisted of a probably representative twelve individuals: six adults, three adolescents, two juveniles, and an infant. DNA diversity was quite low among them all, but the three adult males were very closely related, while the three adult females came from other DNA lineages, suggesting that the core Neanderthal group was composed of males, while females left at puberty to live with other groups.

The Neanderthals employed flexible economic strategies that varied according to what was available in the local habitat, and they clearly understood and exploited their environments with great subtlety and finesse. Sometimes, small as their groups might have been, they hunted fearsomely large mammal prey such as mammoths and woolly rhinoceroses, using stone-tipped spears. At other times and in other places they went after much smaller game—at one Mediterranean site they apparently specialized in tortoises. And everywhere, they exploited whatever plant resources were available—though the colder it got in those times of notoriously unstable climates, the greater their dependence would necessarily have been on those big grazing mammals.

Shortly before their disappearance, the Neanderthals were joined in their European heartland by modern humans, African in origin, who had begun to trickle into the region by about forty-three thousand years ago. There is evidence from DNA (not entirely uncontested) of some genetic intermixing between the resident hominids and the newcomers, but interaction of this kind would not be unexpected between two young species that had shared a common ancestry not much more than half a million years earlier. What is more important is that such interaction seems not to have materially affected the future biological trajectory of either participant in the process. The Neanderthals went on to become extinct identifiably as the creatures they had been for two hundred millennia; and, the odd "Neanderthal gene" notwithstanding, we are basically the same species today as we were then.

The precise reasons for Neanderthal disappearance have been much debated: Were they assimilated? Exterminated? Outcompeted? Infected with pathogens? Were they already on their way out, and simply vulnerable to a final minor coup de grâce? Anything and everything is possible, and a recent report does appear to document pretty conclusively that the Neanderthals were under particular climatic stress around the time of their disappearance. But, irrespective of the details, there can be little doubt that in a global sense the arrival of *Homo sapiens* was implicated in the Neanderthal extinction. As, indeed, modern human incursion almost certainly was in the broadly contemporaneous disappearances of *Homo erectus* and *Homo floresiensis* in Asia, and of any lingering competition in Africa. *Homo sapiens* was literally the only new thing in the experience of any of those formerly hardy survivors. And its victory in this dramatic process of triage among the hominids—which made it, for the first time since the very beginning, the lone member of its family on the planet—tells us that there was something entirely unprecedented about the newcomer.

Looking back over the Neanderthals' record, it is hard not to be impressed by what those hominids achieved. But, significantly, what we don't find in that record is any plausible sign that *Homo neanderthalensis* was regularly thinking in a symbolic manner. Certainly, the Neanderthals were highly skilled and intelligent; but, in contrast to *H. sapiens* they left us precious little reason to believe that, as a group, they had qualitatively broken with the past in the way in which they used information. Quite possibly, they contrived to take the ancestral intuitive cognitive algorithm to its ultimate limits; but in the end, they seem to have been employing those big, hungry brains simply to do, if a little bit better, basically the same things their predecessors had done with theirs. This means that it is highly misleading—and fails to do the Neanderthals proper justice—to view them simply as a failed version of ourselves.

None of this is to deny that the Neanderthals were complex creatures, capable of some very sophisticated behaviors. At one quite early Neanderthal site, for example, some eagle talons were found that had apparently been grooved for stringing; and at a very late site, a hash-like sign was deeply engraved into a cave floor covered with Mousterian deposits. There has even been a recent suggestion (not uncontested) that Neanderthals had dabbed pigment on Spanish cave walls before *Homo sapiens* arrived in the area. But, once more, these are individual expressions, devoid of any larger social context that might support the idea that Neanderthals in general were manipulating information in the modern human symbolic manner.

The notion is occasionally tossed around that some acculturation may have occurred during the time that Neanderthals and modern humans overlapped in Europe, with the implication that the two species were on a par in terms of processing cultural information. However, most putative evidence for acculturation comes from a mysterious late culture called the Châtelperronian, in which any meaningful association of Neanderthals with limited Cro-Magnon-like expressions is appearing steadily less certain. Of course, if as the molecules suggest there was a limited amount of genetic exchange between *Homo sapiens* and *H. neanderthalensis*, some kind of interindividual and possibly intercultural interaction must have occurred between the two. But exactly what form that interaction might have taken still entirely eludes us.

At present, then, it seems overwhelmingly probable that, as smart as the Neanderthals may have been, and as much as they clearly had in common with us, they were their own unique entity. They were not simply an alternative version of *Homo sapiens*, either phylogenetically or cognitively. After all, *Homo sapiens* has contrived to transform the planet in a geological eyeblink; and in the substantial span of time and space they inhabited, if the Neanderthals had reasoned and acted in the way we do, they would surely have left us some more substantial evidence of the fact.

ORIGIN OF *HOMO SAPIENS*

The very first indications we have that creatures who looked just like us were up and about on the planet come from sites in northeastern and southern Ethiopia that date between about 200,000 and 160,000 years ago. The roots of our species may well go deeper yet, of course, but currently all reports purporting to substantiate this are unreliable. What's more, at this point we do not have any fossils older than 200,000 years that even convincingly anticipate our curious cranial construction, with a small face retracted beneath the front of a delicate globular braincase, or the remarkable lightness of build of our body skeletons. Indeed, we know substantially more about the history of the Neanderthals, who have identifiable predecessors going back in the fossil record to over 400,000 years, than we do about the emergence of our own species. Partly, this may be a simple issue of discovery: our lineage evolved in Africa, a vast continent that so far has been relatively poorly explored by paleontologists, while the Neanderthals and their predecessors flourished in a compact peninsula of Eurasia that has been the backyard of antiquarians and archaeologists for centuries. But the fact is that, right now, it remains obscure exactly where in a dimly glimpsed record our immediate precursors sprang from.

There is one thing we do know, however. Namely, that the very earliest anatomical representatives of *Homo sapiens* in Africa are accompanied by an unremarkable archaeological record. So far, this record has yielded no convincing evidence that those very early modern humans were behaving significantly differently from, for example, their much-better-documented Neanderthal contemporaries farther north. Indeed, the 160,000-year-old northeastern Ethiopian locality of Herto, where the most complete of the early *Homo sapiens* fossil specimens were found, also boasts the very latest handaxes ever found in situ on the African continent: an unexpectedly archaic material feature at that late date. Once again, the advent of a new hominid species had apparently failed to coincide with any cultural innovation at all, let alone any indication that it was thinking or behaving symbolically.

For such indications we are obliged to wait for several tens of millennia after the first appearance of anatomical *Homo sapiens*. Beginning at a little over one hundred thousand years ago, presumed *Homo sapiens* sites on the fringes of the Mediterranean and in southern Africa start yielding evidence of shell beads that were pierced for stringing. Often these are found in association with ground ochre, a material that has strictly practical uses but that in historic times has also been widely associated with bodily decoration. Among modern humans, personal adornment is typically loaded with symbolic significance, making statements about the wearer and his or her economic or social status; and while not everyone would accept individual ornamentation on its own as definitive proof of symbolic cognition on the part of any prehistoric people who practiced it, given the larger context it is nonetheless very strongly suggestive in this case. Fortunately, though, more concrete indications that people were becoming symbolic were soon forthcoming.

Blombos Cave, on the southern African coast, was sporadically occupied by early *Homo sapiens* between about one hundred thousand and seventy thousand years ago. At this site archaeologists have found not only evidence of shell beadwork, bone tools, microliths (tiny stone blades set into handles), and elaborate ochre processing, but also some small, smoothed ochre plaques bearing deliberately engraved geometric designs. The most striking of these are about seventy-seven thousand years old, but others are earlier; they are also accompanied by the world's earliest known drawing: some hashed lines sketched on a silcrete pebble some seventy-three thousand years ago, using an ochre crayon. In the aggregate these objects suggest a symbolic sensibility and a continuity of meanings conveyed and understood across the generations. This probability receives further support from colored and geometrically engraved ostrich-eggshell fragments known from not very far away at the slightly later Diepkloof Cave.

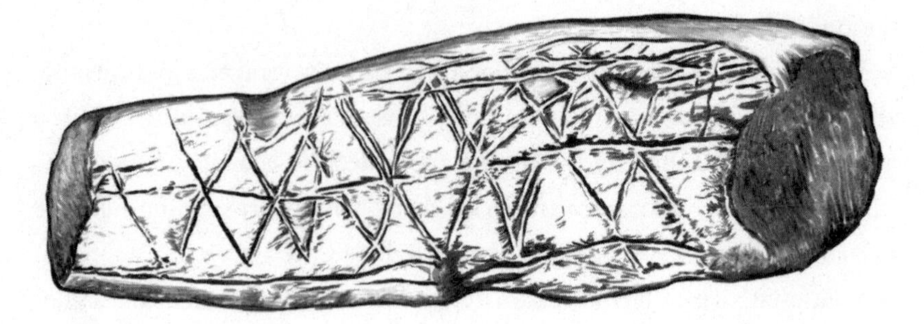

Smoothed and engraved ochre plaque from South Africa's Blombos Cave. At around seventy-seven thousand years old this is the earliest explicitly symbolic product of *Homo sapiens* yet found, although earlier indirect evidence is provided by pierced marine gastropod shells. *Drawn by Kayla Younkin.*

The cultural context in which these unprecedented objects appear is known to African archaeologists as the Middle Stone Age (MSA). Until not long ago, the MSA was regarded as the roughly contemporaneous African equivalent of the Middle Paleolithic cultures of Europe, among which the best known is the Mousterian of the Neanderthals. But in recent years it has become ever clearer that something was beginning to stir in Africa during the MSA that found no equivalent in the European Middle Paleolithic.

Aside from the geometric symbols we've just mentioned, we see a qualitative leap in technology at both Blombos and at a cave complex called Pinnacle Point, a little way along the coast. In an area where good toolmaking stone was scarce, MSA people at both sites employed a sophisticated method of heat-treatment to convert a soil derivative called silcrete from a mediocre toolmaking material into an excellent one. At Pinnacle Point, which also yielded abundant evidence for pigment processing, this technology was flourishing by about seventy-two thousand years ago, and it may have initially appeared considerably earlier. Our general belief is that most Paleolithic technologies and other purely functionally directed activities, however sophisticated and skilled, cannot be taken on their own as proxies for symbolic thought. There are, after all, other ways than our own to be smart and technologically adroit. But, because of its sheer complexity and need for forward planning, the multistage heat-treatment of silcrete is almost certainly an exception, as is a Paleolithic ceramic technology documented considerably later in the Czech Republic. What is more, the archaeologist Curtis Marean has strenuously argued that the intensive shellfish-based economy of the ancient inhabitants of Pinnacle Point was one that required a capacity for predicting tides and other natural phenomena that would have been impossible without sophisticated cognition of the modern human kind.

So, clearly, something significant was happening in the later MSA in the period between about one hundred thousand and seventy thousand years ago: something that had no precedent in the archaeological record, either inside or outside Africa. Significantly, this date range places those developments just before the point at which molecular anthropologists estimate, from DNA data, that modern humans left Africa and remarkably rapidly took over the Old World, displacing all resident local species of *Homo* in the process. There is evidence from China that *Homo sapiens* had arrived by perhaps as much as seventy thousand years ago, and by not long thereafter humans had crossed fifty miles of sea to reach Australia. By shortly after forty thousand years ago, *Homo neanderthalensis* was gone in Europe and western Asia (as evidently were the Denisovans, Neanderthal relatives basically known only from their DNA). *Homo erectus* and the odd "hobbit" *Homo floresiensis* were gone from eastern Asia. And we have no evidence later than this time for the persistence of any other *Homo* species in Africa.

Because the pattern previously had been for multiple hominid species to coexist, even on limited areas of landscape, this story of rapid replacement everywhere suggests by itself that there was something significantly novel about these early *Homo sapiens* migrants. They were relating to their surroundings, including their closest relatives, in an entirely new way that pointed to significant cognitive novelty. In contrast, the presymbolic humans of modern anatomy who contrived to penetrate the Levant by well over one hundred thousand years ago had never come even close to edging out the resident Neanderthals.

At this point, you may be wondering about the reports published over the last few years of both DNA and fossil indications of interbreeding among the African émigrés and the hominid populations they encountered on their forays north and east. The fossil evidence bearing on this question is equivocal, and the DNA conclusion is somewhat algorithm dependent; but in fact, as we have already intimated, some intermingling between members of recently diverged young species would be expected. More important is that any interbreeding which may have occurred does not appear to have had biologically significant consequences. The Neanderthals, for example, went on (with alarming rapidity) to become extinct very much as themselves, while (the odd "Neanderthal gene" aside) *Homo sapiens* equally retained its identity.

The material record of the exodus from Africa remains a bit scanty; and although as a result it is still debated whether there was one exodus or many, it is almost certain that we are looking at an ongoing process rather than at a single event. Still, irrespective of just how many waves of emigrants left Africa in the period centering on seventy thousand years ago, one thing is certain: they were all behaving symbolically. They had all acquired the cognitive hallmarks that distinguish people today.

There are ample lines of evidence for this, but none is more striking than the outbursts of artistic expression that we see in both Europe and eastern Asia at around the forty-thousand-year mark. There can be no more dramatic illustration of the symbolic gift, no more profound demonstration that what has been called "the human capacity" had arrived, than the creation of figurative (and concurrently highly symbolic) art—especially if, as in Europe, it is accompanied by abundant evidence for a much more complex lifestyle and economic strategy than anything known earlier. In France and Spain the richly documented era of Paleolithic cave art, which occurred between about forty thousand and ten thousand years ago, produced some of the most powerful artistic expressions ever created; and the meticulously observed and often exquisitely rendered animal images of the period are accompanied by a profusion of geometrical and other signs that clearly also bore symbolic meaning. Without knowing the appropriate cultural referents we cannot know what those signs meant, or why those who made them frequently penetrated extremely difficult and uncomfortable spaces to carry out their work; but in those messages from that most distant of human pasts, we unquestionably see the workings of the complex and frequently unfathomable modern human spirit.

Monochrome rendering of a now badly faded polychrome wall painting from the cave of Font de Gaume, France. A female reindeer kneels before a male, who is leaning forward and delicately licking her brow. *Drawn by Diana Salles, based on an original rendering by the Abbé Breuil.*

Ancient representational art was not a uniquely European phenomenon. This was confirmed recently by the dating of some hand stencils and animal representations on the walls of a rock shelter on the Indonesian island of Sulawesi that determined they were made about forty thousand years ago. Indeed, this discovery (and a similar one announced in Borneo as this book goes to the printer) strongly suggests that representational art was not a European invention at all, but rather that the twin traditions in northern Europe and eastern Asia both derived from an earlier practice, most plausibly invented in the parent continent of Africa. And if this is the case, we can reasonably infer that artistic activity, and all the other behaviors for which we can regard it as a proxy, followed very soon after modern human cognition was acquired there. Still, whether representational art was invented independently in Europe and Asia, or descended in both places from a common African source, there can be little doubt that the capacity that underlies such activity was already in place when the first émigrés of the modern period left their natal continent.

BECOMING HUMAN

The African MSA is thus the material witness to what was certainly the most extraordinary behavioral revolution ever to come about within a living species. Indeed, nothing that had ever happened before could have predicted that a smart but entirely intuitive species would ever become transformed into a symbolically reasoning one. After all, the two types of cognitive system are entirely discontinuous: the difference between them is a qualitative one, not merely a more-is-better quantitative one. Besides, the only reason we have for believing that such a transformation *could* ever have occurred, is that it evidently *did*. And it happened so fast that there is no way we can implicate the long-term workings of natural selection. So, what on earth was going on as humans adopted symbolic reasoning?

Well, as we have already seen, there is nothing unusual about behavioral innovations being made within species. In fact, that is where they *must* happen. There really is nowhere else. Equally importantly, before a new behavior is adopted the structures that permit it must obviously be in place, because without them it could never be expressed. This in turn means that the organism in question needs to actively discover that the novel structures concerned can be used for a new purpose. And the agent of that discovery—should it ever be made—must necessarily be a behavioral one, given that the enabling biology must already be there.

We have already pointed out that science is only at a very rudimentary stage of understanding what it is that generates our individual consciousnesses; but in the case of symbolic reasoning, we can reasonably surmise that the vital biological underpinnings lie in nervous pathways within the brain that make possible direct associations between the inputs and outputs of diverse brain regions. As physical structures, these pathways must have had their origin in a biological event; and the most obvious candidate for the event concerned is the one that gave rise, some two hundred thousand years ago, to *Homo sapiens* as a distinctive anatomical entity.

As we have intimated, we do not have any useful fossil record that would help us understand exactly what happened in this fateful episode of speciation, although we do know that it involved a radical developmental reorganization that affected the skeleton—which is all that is preserved in the record—from head to toe. But the fact that we have only the skeleton gives us no reason to believe that the event concerned didn't also affect the soft tissues of the body, including the brain itself. In which case, the most likely scenario is that after several hundred million years of accretionary vertebrate, primate, and hominid brain evolution, a brain structure was acquired in this event that adventitiously possessed the necessary physical structure for symbolic thinking.

Some authorities have suggested that the symbolic brain resulted from an otherwise undetectable genetic innovation that was acquired at the point at which humans started behaving in the modern manner. But there is no way of demonstrating this; and, as we have just suggested, it is much more likely that it was attained in the event of extensive developmental modification that produced the idiosyncratic physical structure of *Homo sapiens*. The new cognitive potential then lay fallow, as it were, until it was "discovered" through the action of a behavioral stimulus. The result of this discovery was, of course, extraordinary; but as we saw in our earlier discussion of exaptation, in evolutionary terms there need have been nothing remarkable about the process itself.

So, what might that behavioral stimulus to symbolic thought have been? For several decades now, the driver of modern human intellectual complexity favored by many students of human behavioral evolution has been theory of mind—basically, the ability of one individual to assess what is going on in the mind of another. The argument goes that higher primates such as the apes and us are highly social creatures, and that in a complex society an enhanced ability on the part of one individual to read the intentions of another, and thereby to predict its behavior, will put the mind-reader at a competitive premium. Because higher primate societies are to a remarkable degree about male access to females, this premium will be expressed in reproductive success; and if the superior cognitive qualities at issue are of the inherited variety, over the generations those qualities will become more widespread in the population, which will as a result become on average smarter.

In the human case, it is often assumed that a mechanism of this kind accounts for the remarkable increase in brain size that we see in the genus *Homo* over the past two million years or so (because surely, bigger must equal more intelligent, both socially and—more visibly—technologically). However, it is not at all obvious why a general effect of this kind should be operative only in the genus *Homo*, rather than among all lineages of highly social higher primates; and it is, moreover, a prime example of the kind of gradual mechanism that cannot possibly account for a sudden emergence such as we see in the case of modern human symbolic behavior.

Theory of mind becomes an even less attractive mechanism for propelling hominid brain size increase when we consider that modern apes are, in fact, quite good at reading the minds of others. It has recently been shown that an individual ape is capable of understanding that others may harbor beliefs that it (the observing ape) knows to be false. That is pretty impressive, and it is something that a three-year-old human child cannot yet do. But there is still a crucial difference here between adult apes and humans, because apes understand the false-belief situation on an intuitive, implicit level (showing purely in its reactions that it possesses that understanding), while humans comprehend it on an explicit level (and can convey it to others). It is our very recently acquired symbolic cognitive style that permits our knowledge to be made explicit (both to ourselves and to others) in this way; and this cognitive mode appears to be independent in origin from the basic capacity for theory of mind, even though it is probably what permitted the latter to become more explicit among humans and thus more complex (one individual can understand that another may have a different belief that is rejected by a third individual but accepted by a fourth, and so on).

It is very important to emphasize here that, while the strong but metabolically costly tendency to enlarge brain size over time that we see in the genus *Homo* is undoubtedly linked in some way to increasing "intelligence," it did not inexorably lead toward intelligence of the symbolic kind. This is made clear by the simple observation that brain size increase in the history of our genus proceeded independently in several different lineages, only one of which achieved symbolic thought—or, at least, displayed indisputably symbolic behaviors. We see brain enlargement over time in the lineage that led to the emergence of our own ultimately symbolic species in Africa. But we also see it in eastern Asia, where earlier *Homo erectus* had smaller brains than later ones. And we see it in Europe, too, where early relatives of *Homo neanderthalensis* had smaller brains than those possessed by the later descendants that survived to encounter the Cro-Magnons (whose brains were of virtually identical size to their own, even if apparently organized differently).

The bottom line thus appears to be that, since modern symbolic cognitive processes appear to be unique to later *Homo sapiens*, the stimulus that drove them into existence cannot have been the driver of the trend toward raw brain size enlargement that embraced the genus *Homo* as a whole. That trend was underpinned by something else, most likely having to do with the increasing refinement of the ancestral intuitive mode of intelligence. Whether this was an intra-lineage phenomenon, or whether it was alternatively driven by competition among different hominid species, remains to be determined.

IN THE BEGINNING WAS THE WORD

So just what was it that propelled the abrupt adoption of symbolic cognition by early members of our species—but only after it had already been in existence as an anatomical entity for a geologically short but nonetheless significant length of time? Without doubt, the best candidate by far is the spontaneous invention of spoken language—which is, after all, the ultimate symbolic activity. Exactly like symbolic thought, language depends on the deconstruction of experience into a vocabulary of those discrete mental symbols we wrote of earlier. Moreover, after a period during which linguists tended to compartmentalize language and thought, they are actively beginning to reconsider the wisdom of having done so. As our linguist colleague Wolfram Hinzen has observed, language and thought "are not two independent domains of inquiry." Indeed, they are so intimately intertwined that it is virtually impossible for us to conceive of one independently from the other. They may be conceptually separable, but functionally they are impossible to disentangle.

In addition, we require a stimulus that could have produced a radical result virtually instantaneously; and language scores well here too. The linguists Bob Berwick and Noam Chomsky have recently argued very persuasively that the algorithmic basis for language is very simple. So simple, indeed, that even without knowing what immediately preceded it, we can plausibly conclude language was adopted in a single shift. And linguists studying language development in the field have actually observed a sign language—one with all the structural peculiarities of spoken language—spontaneously emerge and rapidly become more nuanced. This occurred among hearing-impaired but cognitively intact Nicaraguan schoolchildren after they had been brought together for the first time in specialized schools for the deaf—a nice analogy to what may have happened in the origin of spoken language. Once invented, moreover, language would have been ideally poised to spread rapidly through populations of early humans who were already biologically enabled for it—people who had, as some neurolinguists like to put it, "language-ready brains," just like those Nicaraguan kids.

Language was, of course, a radical behavioral innovation, and it is one to which adults might have been at least initially resistant. But it is not hard to imagine, at least in principle, how some early *Homo sapiens* children, living in some small isolated population in Africa during a time of environmental stress, might have spontaneously started attaching meaning to sounds they already employed in the context of a sophisticated prelinguistic vocal communication system. These associations might have started a feedback between sound, symbol, and meaning in their language-ready brains, leading not just to an organized communication system but, simultaneously, to structured thought patterns. Both of these would then have been open to later mutual refinement; and the new behavior might have subsequently spread upward through the social hierarchy, in much the same way that the practice of washing grit off sweet potatoes before eating them famously spread upward through the dominance hierarchy of a group of Japanese macaques. In offering this rather simpleminded scenario we hope we do not appear to be trivializing what was undoubtedly the most profound cognitive transition that ever occurred in the history of life; but it is overwhelmingly probable that, however extraordinary the effect of this transition, it ultimately came about through the operation of a relatively mundane process.

Interestingly, the notion that modern cognition emerged via a sudden alteration in the brain's algorithmic functioning is supported by the rather surprising observation that, following some two million years of steady expansion, the average brain size of *Homo sapiens* has plunged since the late Paleolithic. At almost exactly 1,500 ml, the mean endocranial volume of a sample of twenty-nine Ice Age *Homo sapiens* skulls is virtually identical to that of a similar sample of Neanderthals, at 1,487 ml. Yet today, the mean volume of *Homo sapiens* cranial vaults is a mere 1,330 ml—a decrease of almost 13 percent. Various imaginative scenarios have been put forward to explain this rather counterintuitive phenomenon, but the most straightforward possibility is that the symbolic brain is simply working on a more metabolically frugal algorithm than its intuitive predecessor. We have seen that maintaining brain tissue is an expensive proposition, and that no species is likely to maintain any more of it than it absolutely needs. The most parsimonious explanation for the apparently inexorable Pleistocene increase in average hominid brain sizes is thus that, under the ancestral intuitive algorithm, "intelligence" (however you want to define this tricky quality) scaled more or less directly with raw brain volume. If it was an advantage to be smarter, you needed a bigger brain. Toward the end of the Pleistocene, however, the *Homo sapiens* brain switched to the new and more energetically economical symbolic algorithm: an algorithm that (entirely adventitiously) produced a competitively superior cognitive product while necessitating a smaller quantity of raw brain tissue.

Whatever the exact mechanism involved, the evident abruptness with which modern human cognition emerged suggests something that is of fundamental importance when we try to understand our human behaviors today. For the new symbolic capacity was grafted on to what we have been calling the "intuitive mind," which incorporated some extremely ancient cognitive processes. History has ensured that the ape mind still lurks inside our heads, with yet older systems inside it. But although the strong imprint of that history remains—for we are far from entirely rational beings—the overlayering of the ancestral brain with the associative apparatus that permits the symbolic capacity changed all the rules. And, what's more, it did so very quickly indeed. The net result is that, although our brains and minds are unquestionably the products of an enormously long history of gradual accretion over evolutionary time, we cannot view our behaviors as having been fine-tuned by natural selection. We are not, as some suppose, the prisoners of a hypothetical Ice Age "environment of evolutionary adaptation" that reflects itself in bizarre and inappropriate reactions to our changed modern circumstances.

Instead, for all that they are often mediated by some very ancient systems, our cognitive processes are both emergent and nonspecific, endowing us with the basis for our free will—no matter how much that free will may be compromised in practice by both intrinsic and extrinsic circumstances. Through extended periods of infancy, juvenility, and adolescence, we modern *Homo sapiens* learn an unprecedented amount of behavior from our elders and peers. This indoctrination will inevitably affect how we behave in our later lives, as will our later circumstances. Still, once adult, we have an entirely unmatched choice in the ways in which we behave as individuals. Put simply, because we can *imagine* more choices than our binary brethren can, we can *make* more choices. And one consequence of this is that simply knowing that someone is a member of *Homo sapiens* predicts surprisingly little about him or her, apart from the fact that he or she is a biped—with all the associated physical consequences—and processes information in an unusual way.

All this having been said, it remains true that every living human being belongs to the same species; and we are all, so to speak, in the same evolutionary boat. Behaviorally flexible though we may be, we all share something we can recognize as a common humanity: an extraordinary quality that we have acquired through routine evolutionary processes. In the remainder of this book we will explore whether that intuitive notion of common humanity can be distilled into something we can usefully label the human condition.

FOUR

GENES, PEOPLE, AND BEHAVIOR

By any biological criterion you might want to consider, all human beings on the Earth today belong to the same species. But although any one of us is capable as an infant of learning any language, or of absorbing the customs and values of any society, by the time we are five years old, we have already learned a set of values and beliefs—and prejudices—that we will struggle to free ourselves from later in life. Ironically, much of this learning takes place during the period of "childhood amnesia" from which we will later find it hard to recall any memories. As a result, we will never be consciously aware of how we acquired many of the attitudes and assumptions that will subsequently govern our interactions with the world. In the most extreme cases, by the time they have both reached adulthood a member of one culture will see the world entirely differently from a contemporary brought up in another; and in a highly specific and limited sense, the two might as well belong to different species, divided by language, assumptions, priorities, and even by different ideas of right and wrong. Those differences might not matter too much if the individuals involved were confined to opposite sides of the planet; but in a rapidly globalizing world the inherent potential for conflict is obvious. You could not say anything even remotely similar about members of any other species on Earth; and, once again, the difference resolves down to our individualized ways of reformulating the world in our heads. And while this idiosyncrasy of ours hardly suggests that our patterns of behavior are biologically ingrained, that inconvenient fact has not prevented a whole area of science from being founded on the assumption that they are.

SOCIOBIOLOGY

As we noted earlier, in 1975 a highly anticipated and widely read book called *Sociobiology: The New Synthesis* was published by Harvard Press. A magisterial survey of behavior across all groups of social animals, it was written by E. O. Wilson, an immensely accomplished entomologist and biodiversity expert specializing in the highly social ants. *Sociobiology* was written with the authority of a textbook, and indeed it was used as a text in many animal behavior courses, serving its principal purpose well. Organized into three main sections, it outlined a strong research program for studying animal behavior. It started with the principles of evolutionary analysis, then went into the mechanisms of social interactions, and ended with a comprehensive survey of social organisms in the natural world. There was only one problem with the book: its Chapter 27. This chapter outlined a research program using behavior in other animals as a springboard to study that of humans. Wilson felt that because his ants were social, some of the generalizations he had derived from them could be applied to undoubtedly social humans. Indeed, Wilson went so far as to claim that "the systematic study of the biological basis of all social behavior" could be carried out using his comparative approach. Note that he used the word "all" in relation to social behavior, and this is where the problems started.

Inevitably, Wilson found himself buried in an avalanche of criticism, particularly from his Harvard colleagues Richard Lewontin and Stephen Jay Gould. His lectures were frequently picketed, and at one he was literally pronounced "all wet" by left-wing protesters who dumped a pitcher of cold water on his head. To the extent that ants are entirely constrained in their behaviors in ways in which human beings are signally not, the protesters had a point, whether or not you might approve of the way they made it. Still, there is evidently no such thing as bad publicity, and Wilson and his sociobiology became a hot-button item for some of those interested in the evolution of behavior. Indeed, several journals (*Sociobiology*, *Ethology and Sociobiology*, and *Behavioral Ecology and Sociobiology* among them) were initiated to promote and publish work in the area. Rarely has a truly scholarly book provoked such strong and polarized reactions, and clearly it was Chapter 27 that was principally at issue.

A couple of years earlier Michael Ghiselin, a philosopher at the California Academy of Sciences, had very quietly coined the term "evolutionary psychology" for one particular offshoot of Darwinian thinking. Using Charles Darwin's 1872 book *The Expression of the Emotions in Man and Animals* as a stepping-off point, he described a research program in human psychology that had its roots directly in evolutionary biology. This was a perfectly legitimate thing to do, especially since Darwin himself had predicted, as early as 1859 in the *On the Origin of Species*, that "in the distant future . . . psychology will be based on a new foundation, that of the necessary acquirement of each mental power and capacity by gradation." The great evolutionist had a grand vision of nature, yet he always remained painfully aware of the idiosyncrasies and complexities of the living world. But in contrast, the evolutionary thinking subsequently used to construct evolutionary psychology in his name was notably narrow and reductionist.

The principles of the burgeoning field of evolutionary psychology were clearly articulated by the University of California, Santa Barbara, psychologist Leda Cosmides and her anthropologist husband, John Tooby, in a series of papers published during the 1980s. Their goal was to unite cognitive psychology with evolutionary biology. For this union to make sense, they had to dismantle a basic organizing principle of the social sciences: what they called the SSSM (the standard social sciences model). As they describe it, this model maintains that "the contents of human minds are primarily (or entirely) free social constructions, and the social sciences are autonomous and disconnected from any evolutionary or psychological foundation." That cultural description matches our messy cognitive processes pretty well, but Cosmides and Tooby wanted to reduce the problem of human cognition to a strictly reductionist biological one.

All of this may sound a lot like sociobiology, but evolutionary psychologists claim otherwise. In fact, the word "sociobiology" does not occur even once in Cosmides and Tooby's online *Primer on Evolutionary Psychology*. As a result, several new journals had to be started to accommodate the burgeoning evolutionary psychological research program (*Evolutionary Psychology*; *Evolutionary Behavioral Sciences*; *Journal of Evolutionary Psychology*, *Evolution, and Human Behavior*; *Journal of Social, Evolutionary, and Cultural Psychology*). Still, if you look closely at the goals and approaches of evolutionary psychology and sociobiology, it is impossible to miss an intimate intellectual connection between the two. And indeed, sociobiology itself recognizes that connection, not least in Harvard University Press's own online blurb for the twenty-fifth-anniversary edition of Wilson's now-classic book: "Human sociobiology, now often called evolutionary psychology, has in the last quarter of a century emerged as its own field of study, drawing on theory and data from both biology and the social sciences."

So, let's look a little more closely at the connection. In their *Primer on Evolutionary Psychology*, Cosmides and Tooby concluded that:

> All normal human minds reliably develop a standard collection of reasoning and regulatory circuits that are functionally specialized and, frequently, domain-specific. These circuits organize the way we interpret our experiences, inject certain recurrent concepts and motivations into our mental life, and provide universal frames of meaning that allow us to understand the actions and intentions of others.

In other words, Cosmides and Tooby were advocating replacement of the SSSM by the following five principles:

1. The brain is a physical system. It functions like a computer.
2. Our neural circuits were designed by natural selection to solve problems that our ancestors faced during our species' evolutionary history.
3. Consciousness is just the tip of the iceberg; most of what goes on in your mind is hidden from you.
4. Different neural circuits are specialized for solving different adaptive problems.
5. Our modern skulls house a Stone Age mind.

Both their definition of evolutionary psychology, and their five-pronged approach to it, point to the kind of reductionism that was also typical of sociobiology as applied to humans.

Another weakness that evolutionary psychology shares with sociobiology is a reliance on adaptation as a linchpin of the explanatory process. We have criticized the adaptationist paradigm earlier in this book, as Richard Lewontin and Stephen Jay Gould also did, very comprehensively, in their criticisms of sociobiology in the 1970s. And it is appropriate to point out also that, while it is absolutely correct to say that our brains are physical systems, it is much less accurate to compare them to computers, as sociobiologists so often do. For one thing, the computers we used to write these words on did not evolve in the way our brains did. As our Apple laptops gradually became what they are today, earlier design elements were unceremoniously discarded, and entirely new starting points were adopted. The existing variation in earlier computers like the Apple II was not used in the development of subsequent Apple products; only the best design elements were carried forward. And, far from being the products of a random mutational process, the novelties in the new designs were deliberately conjured up by the human imagination.

KLUGES AND TEAMS OF RIVALS

To claim that our brains are like computers implies a striving toward perfection that we hope we have already convinced you does not exist in the evolutionary process. And indeed, we deeply believe that the human brain is something no engineer would (or even could) ever design. Engineers optimize, whereas our brains are optimized for nothing—which is, quite plausibly, exactly why they are such flexible and innovative mechanisms. Our colleague Gary Marcus has aptly described the human brain as a "kluge," a Rube Goldberg–style apparatus that doesn't make much sense to engineers because of its haphazard design, but which turns out to work quite well. The neurobiologist David Eagleman, author of *Incognito*, describes our brains as "teams of rivals," recalling the term used by the historian Doris Kearns Goodwin to characterize the strategy used by Abraham Lincoln in assembling his cabinet during the Civil War. Lincoln's cabinet harnessed opposing views and impulses into a single working entity. Similarly, our brains are better described as high-functioning messes than as computers.

Reducing the evolution of our neural circuits to an agglomeration of discrete units that might be under selective pressure is actually a massive oversimplification of how our brains and nervous systems work, and this ugly fact also fatally undermines the equally unsettling notion that we have a "Stone Age mind" that is necessarily frozen in the past because natural selection has not yet had time to catch up with a fast-changing world. Of course, at some level the evolutionary psychologists must be right, in that our brains are somehow "adapted" to what we do, at least in the sense that they produce behaviors that are (so far) compatible with our continued existence in the huge array of environmental circumstances in which we find ourselves. But this is a far cry from saying that they have been specifically programmed by nature to generate those behaviors. Our brains are general-purpose organs that are there to aid our species in survival and to help the members of a highly social species deal with each other. These functions involve integrating and dealing with numerous conflicting signals, and it may be worth taking a moment to look at how the human brain deals with conflict.

How people behave as a result of certain brain injuries, or of surgical intervention to treat them, can illustrate how clever our brains are when dealing with conflicting information. A particularly intriguing instance of how far our brains will go to make sense out of conflicting information received from the senses is furnished by people with disrupted communication between the two hemispheres of their brains. "Split brain" individuals of this kind have been studied following intentional surgery aimed at alleviating severe cases of epilepsy. When the corpus callosum (the region of the brain that connects the right and left hemispheres) was surgically split in this cause, the neural circuitry between the hemispheres was severed, and the epileptic fits stopped. That much was fine; but since our left and right brains need to communicate with each other to properly interpret the information coming in from the outside world, one deficit was exchanged for another.

The surgery was done for clinical reasons, but the neuroscientists involved also spotted an opportunity. By setting up a system whereby the left and right eyes viewed different things, and then asking questions about what the eyes saw, they could discover amazing intricacies in how their patients' brains dealt with visual noise to produce what the psychologists call a "unified sense of self and mental life." One of the more famous experiments of this kind concerned flashing the word "face" to the left eye, and the word "smile" to the right eye of a patient, who was asked to describe what he or she had seen by making a drawing, and then to explain the drawing verbally. In one famous case, a patient was asked to draw what he had seen, using his right hand (which is controlled by the left brain). He drew a smiling face. So far so good. But when asked to describe in words why he had drawn the face smiling the patient, who couldn't integrate the information from the separate sides of his brain, had no idea. But he still felt the need to confect an explanation that sounded good, so he answered that he just drew a face, and that a smiling face is more pleasant than a frowning one is. "Who wants a frowning face around?" he asked. To anyone with normal communication between the two brain hemispheres this response might sound odd, to say the least. But it is completely explicable in terms of right brain/left brain dynamics. Having limited information, the patient's left brain made up a logically pleasing story to compensate not only for its lack of information, but also for the haunting need to draw the face with a smile.

Still, by far the most bizarre case of a human brain trying to make sense of sensory noise was the one reported by V. S. Ramachandran and colleagues. Ramachandran, who is most famous for his work on synesthesia and phantom limbs (two other equally compelling phenomena), used the example of Capgras syndrome to demonstrate just how much our storytelling brains want to create explanations for inexplicable sensory input. Individuals with Capgras syndrome will claim that people close to them are impostors. In the case that Ramachandran studied, a man claimed that his mother was an impostor. On being introduced to her, he would say, "She looks like my mom, but she isn't." The emotional response of this individual was measured when he was confronted with his mother, and the analysis indicated that he was responding entirely neutrally to his mother—every mother's nightmare!

Ramachandran offered the following explanation for this bizarre behavior. His subject was an individual who had suffered head trauma. This had damaged his limbic system, in particular the amygdala, a region lying deep inside the brain that is responsible for emotions. Moreover, it is more than likely that the connections of his temporal cortex to the limbic system had additionally been altered in the accident. Also connected to the limbic system is a region of the temporal lobe called the fusiform gyrus. This structure is responsible for processing facial images, and hence for facial recognition. Ramachandran's subject saw his mother, recognized her as such, and tried to send this information to his amygdala for emotional processing. This transfer was thwarted by the damage, leaving the patient with the only logical explanation possible: "This isn't my mother, because I lack emotion for her." Trying to make sense of the nonsensical observations, his brain then made up the impostor story. The clincher for this explanation is that, when his mother spoke to him on the telephone, he recognized her voice and properly sent this information to his amygdala. Hence an appropriate emotional response, and no more impostor story.

While both the split-brain and Capgras syndrome patients are isolated neurological cases, they tell us a lot about how flexible our brains are when reacting to challenging or conflicting situations. The general conclusion has to be that we are not programmed to react directly to incoming sensory information; instead, we use our minds at a more abstract level to join up the dots and find solutions to conflicting or incomplete information. And we would venture that this is true for a very great deal of our behavior: that a lot of what we do or say is the result of similar attempts by the brain to fill in gaps. Indeed, it appears that much of our day-to-day conduct is based on resolving the conflicting information with which our brains are bombarded. We each do this in a very idiosyncratic way, and we strongly doubt that hard-wired adaptation intrudes very much.

The neural phenomena involved in how we sense the world are relevant here particularly because the raw sensory input is transformed into perception, which is then somehow tied into our consciousness and into our personal senses of how we fit into the universe. Combining what we know about our unusual behaviors with what we know about our brains makes it pretty clear that we are not biologically programmed to behave in the exact ways we do; instead, rather than being automatic, our behaviors are continually constructed and reconstructed in our minds. Consider those split brain and Capgras syndrome conditions in which, while the outcome can be predicted in the sense that the subject—being human—will make up a story to resolve the conflict, the story itself will depend entirely on the individual.

While our evolutionary history intimately ties us in to the rest of the living world, and our way of dealing with information is uncontestably the product of evolution, there appears to be no parallel among other animals to our human mental processes. Other creatures will make quick decisions to respond to the challenges the world presents to them, but none of them will rationalize or contextualize what they are trying to process. As a result, as far as we know no other animal species is capable of the incredibly broad range of responses that we human beings show to environmental challenges and conflicts. This is not to demean the responses of other animals—which may be very complex indeed—but it is important to remember that there is a qualitative difference between the ways in which we humans and all other animals process information. This makes it even more critical to get the biology right when we are trying to understand how our brains deal with conflict and challenges in nature.

THE GENES DID IT. OR DID THEY?

Older readers may well remember being amazed when they heard that city district supervisor Dan White had been acquitted of the murders of Mayor George Moscone and Supervisor Harvey Milk in San Francisco City Hall on November 27, 1978. There was no doubt that White had deliberately shot his two fellow politicians, but his jury exculpated him of murder (though not of manslaughter) on grounds of diminished mental capacity, believing the defense's submission that he had become pathologically depressed. In support of this contention, the defense lawyers pointed to huge changes in White's lifestyle in the period before the assault: he became alienated from his wife and his friends, he neglected his personal appearance, and he abandoned a rigorous fitness regime in favor of an addiction to junk food, notably sugar-laden Twinkies. The defense did not, as legend now has it, argue that White's murderous actions were directly caused by the sugary confection; instead, it contended that wolfing down Twinkies was symptomatic of the depression that made him incapable of the rational premeditation that a conviction for murder required. But a good story, like a beautiful theory, is often far too good to abandon; and the term "Twinkie defense" has now assumed quasi-legal status as a synonym for any improbable legal strategy.

A similar dynamic applies to our explanations of human behaviors outside the courtroom. We are a storytelling species, and our predilection for "joining up the dots," by its nature a reductionist enterprise, makes us want to look for easily identifiable culprits. And what more satisfying culprit could there be than the genes that, as everybody knows, are the blueprints upon which we are constructed? In the remainder of this chapter we will look briefly at a small but diverse selection of human behaviors with claimed genetic backgrounds, with the aim of seeing if any such claims have any firm groundings in what we know of the mechanisms of inheritance.

RELIGION

Regardless of whether one has faith or not, one cannot deny that religious belief and its associated effects on social organization have formed a significant part of the fabric of every society ever recorded. Little wonder, then, that it has attracted geneticists and neurobiologists who practice in a new field they call neurotheology. The idea is that, since spiritual feelings both emanate from and are filtered through the brain, they are at least in principle accessible to neuroscience. After all, Francis Crick, quoted earlier, subsequently paraphrased himself by proclaiming specifically that "you, your joys and your sorrows, your memories and your ambitions, your sense of personal identity and free will, are in fact no more than the behavior of a vast assembly of nerve cells and their associated molecules." And there is no reason to think he would have been reluctant to include religious beliefs in this statement as well. We certainly are not. Still, knowing that the brain produces our thoughts and knowing how it does it are entirely different things. If we were somehow to flash freeze the brain of a saint in religious ecstasy, slice it, dice it, and do a rigorous analysis of it, we would be able to recover some biochemical and electrochemical information. But while we would know that this information would in some way be relevant to the individual's state of mind, we would be hard put to say exactly how.

Since Francis Galton did his "effect of intercessional prayer" studies nearly 140 years ago, scientists have periodically revisited transcendental feelings in humans. Not long ago the Templeton Foundation, well known for encouraging work at the intersection of religion and science, funded "The Study of Therapeutic Effects of Intercessory Prayer" (STEP) to the tune of 2.4 million dollars. Grant in hand, Herbert Benson and colleagues contacted three Christian groups and asked them to pray for a cohort of patients undergoing coronary artery bypass grafts (CABGs). A similarly sized cohort was selected but not prayed for. Eighteen hundred patients were followed in all. Alas, prayer was found to have had no impact whatever on the patients' medical outcomes. More strikingly, patients who were informed that they were being prayed for actually developed complications at the rate of 59 percent, while those who were left in the dark came in at 52 percent. The inevitable conclusion was that "intercessory prayer itself had no effect on complication-free recovery from CABG, but certainty of receiving intercessory prayer was associated with a higher incidence of complications," a statement that bears an eerie resemblance to the words Galton penned in 1872: "prayers of the clergy for protection against the perils and dangers of the night, for protection during the day, and for recovery from sickness, appear to be futile in result."

Nevertheless, hope apparently springs eternal about the efficacy of distant healing (prayer), and for reasons that were presumably not all scientific, the National Institutes of Health has just spent nearly two million dollars on funding several studies on the impact of distant healing (prayer) on disease progression and recovery. Whether those studies will reach conclusions any different form Galton's 140 years ago remains to be seen.

In the early 2000s Michael Persinger, a psychologist at Laurentian University in Canada, took the optimism implicit in these awards to an extreme, by employing what later became known as the "God Helmet." This was a device that could focus weak electromagnetic fields on desired areas of the brain to test how stimulation of its various components might affect feelings of spirituality. In the event, Persinger reported that some 80 percent of his subjects responded to stimulation with religious or spiritual feelings, occasionally experiencing self-described full-blown rapture. What he did not claim, however, was to have found a regular, localized "god center" in the brain.

Other scientists have attempted to replicate Persinger's results under rigorous double-blind conditions, but to no avail, and the best guess is that many at least of Persinger's ecstatic subjects were experiencing a "placebo effect." (Believers felt "exceptional experiences" more frequently than non-believers when exposed to a fake God Helmet). Still, Persinger was brave enough to try his device out on the noted atheist Richard Dawkins, the man who described religion as about "turning untested belief into unshakable truth through the power of institutions and the passage of time." Dawkins left the experience with a little dizziness, but with no rapture or visions—and no closer to God.

Still, none of this means that spiritual feelings are not an important part of subjective human experience; and on the social level, organized spirituality can be a strong force for cohesion, especially when accompanied by the ceremony and rituals that really can deeply affect individual experience and mental states under certain circumstances. And of course, religion provides the most venerable of the institutional ways that mankind has devised for soothing the continually nagging and existentially harrowing knowledge of our individual mortality. So given religion's importance, it was inevitable that sooner or later someone would see fit to identify a genetic basis for it. And this was duly done, in 2004, by the notorious gene hunter Dean Hamer in his book *The God Gene: How Faith Is Hardwired into Our Genes.*

According to Hamer, spirituality can be quantified on a "self-transcendence" scale and is to a significant degree heritable. What is more, he fingered the gene called VMAT2, which codes for a protein called vesicular monoamine transporter, as central to the phenomenon. This protein is important in regulating monoamines, chemicals known to be involved in clinical depression. Hamer used a list of nine candidate genes in his study, and on finding that variants of VMAT2 could be associated with transcendental experiences, he leapt to the conclusion he so neatly summarized in the rather extravagant subtitle of his book. The press loved it, of course, although some seasoned scientific correspondents were more skeptical. Carl Zimmer, then writing for *Scientific American*, came up with this lengthy new title for *The God Gene*: *VMAT2—A Gene That Accounts for Less than One Percent of the Variance Found in Scores on Psychological Questionnaires Designed to Measure a Factor Called Self-Transcendence, Which Can Signify Everything from Belonging to the Green Party to Believing in ESP, According to One Unpublished, Unreplicated Study.* This capsule description covers the waterfront for any "gene for," all the way from the low amount of variance explained by single genes in complex traits ("less than one percent") to the difficulties of dissecting the trait (use of "psychological questionnaires") to epistasis ("everything from belonging to the Green Party and believing in ESP") and the frequently nonrepeatable nature of these studies ("according to one unreplicated study").

We have pointed out in other chapters that whenever a claim is made that the "gene for" a particular complex behavior has been found, a red flag should go up. Some geneticists have even come up with an acronym for this approach based on the "one gene, one disorder" principle (namely the highly appropriate OGOD). This is precisely what one should say when hearing about any "gene for" study, because it's just not that simple, for any number of reasons. A more plausible way to think about the complex behavior we call religion is that it may well have arisen independently in disparate societies and cultures, simply based on the unique and innate propensities of the modern human consciousness. History repeats itself because human options are limited, and religion is a very powerful option indeed. As in the case of any independently derived human behavior, the ingredients were quite likely different on each occasion religion was invented, even though the overall principle remained the same—just as the Chinese invented beer based on rice even as the Mesopotamians were inventing beer based on barley. And on the genetic level, the probability that exactly the same gene arrays underwrote the emergence of religion from one ancient culture to the next is actually quite low. In this light, religion might best be seen not as a human universal, but as a complex set of behaviors that share a substrate, but that were independently derived in different cultures.

The underlying capacity for spirituality is, of course, an even more generalized property of humankind than are the religious expressions it makes possible. Spiritual feeling is almost certainly best considered a directly emergent product of the basic human symbolic capacity that allows all of us to imagine the existence of other worlds or other planes of being. It is important to realize that the ability to imagine something is not necessarily synonymous with the propensity to believe in it, which may help to explain the increasing trend toward secularization in rich countries, where current affluence may make any compensating afterlife pale a little in comparison. Whatever the reasons for it, though, the trend clearly cannot be blamed on changing gene frequencies.

ETHICAL BEHAVIORS

In every society, human beings coexist with codes of ethics and morality—whether or not they individually abide by them. And because interindividual and interinstitutional relationships are so complex within societies, it is hardly surprising that recognized norms of conduct have developed to govern them. So fundamental and varied are these norms that in all probability, once human behaviors had become governed by the symbolic cognitive style, they simply emerged as a matter of necessity and subsequently diversified along with the species. Nevertheless, if we cast our net more widely in nature, it is possible to discover among other species senses of fairness and justice that seem to echo those that underpin our own instinctive reactions to what we perceive as appropriate. In this light, it has been argued that our own codes of ethics and morality are built upon more ancient primate, and even vertebrate, foundations.

Back in 2003, for example, experiments with South American capuchin monkeys showed that an individual would reject a piece of cucumber he (or she) would ordinarily have happily eaten if at the same time he saw his neighbor being offered a tastier grape. Clearly, the monkeys perceived that they were being treated unfairly, and felt strongly enough about it to consistently forgo what was, after all, a treat. Subsequent studies revealed similar responses (with different incentives) not only in Old World macaque monkeys and chimpanzees, but in dogs and even (by implication) in rabbits. In reviewing these studies, and others bearing on "mutualistic cooperation" in animals as diverse as elephants, hyenas, and birds, Sarah Brosnan and Frans de Waal of Emory University found that such "inequity aversion" is characteristic of forms that are not only social, but that show cooperation extending beyond the immediate bonds of kinship and mating. In their view, then, inequity aversion is related to the active monitoring by individuals of the cooperative tendencies of others in their social groups, and hence offers evolutionary benefits in terms of social function. And they concluded that "many of the basic emotional reactions and calculations underlying our sense of fairness seem rooted in our primate background."

That a human version of inequity aversion exists can best be demonstrated by a game psychologists call Ultimatum. One player (A) is given a sum of money and is told to split it with the other (B) in the proportions of his choice. A then makes an offer, which B can either accept or refuse. If he refuses, nobody gets anything; but if he accepts, A pays out. In most cases, A will start of by making a reasonable offer to B, say fifty-fifty, and B will almost always accept. If things continue this way, both sides will make a lot of money. But usually A will offer B less as the game goes on. B should accept anyway, since anything is better than nothing, but eventually equity aversion will get the best of him and he will refuse, against the dictates of reason. The golden goose will have been slaughtered. Interestingly, autistic individuals will both offer the worst splits they can get away with and take any offer, apparently because they cannot assess what the other player is feeling or thinking. Both children and chimpanzees, on the other hand, tend to respond like normal human adults. Neural imaging has revealed that the offer to B of an "unfair" deal triggers the insular cortex of his or her brain, a region associated with the emotion of disgust. After the insular cortex has reacted, another brain area called the anterior cingulate cortex kicks in to weigh whether no money or living with disgust is worse, eventually opting for poverty. Such emotional reaction does a lot to explain why the history of humankind is littered with self-destructive decisions and events.

None of this contradicts Brosnan and de Waal's conclusion about the deep origin of our sense of fairness; and of course, much the same thing could be said for many of our other behaviors, including our prosociality and even our ability to make tools. But that is very far from claiming that particular ethical or moral systems have any specific rooting in nature. As a noted geneticist once pointed out, moral norms are the products of cultural, not biological, evolution, even as we exhibit ethical behaviors because our human biological characteristics allow us to anticipate the results of our actions, to make value judgments, and to choose between alternative courses of action. All of these unique human features are properties of a cognitive system that is recent and emergent in origin, rather than having been fine-tuned by natural selection over the eons. Yes, without our primate heritage we would not be what we are. But no, our genes do not dictate our moral or ethical systems. Indeed, it is highly likely that such systems appeared independently in different early societies, as a result of similar functional demands on a common underlying capacity.

WARFARE AND AGGRESSION

While individuals or groups of other organisms may conflict physically, our species is unique both in the breadth and ingenuity of such conflict and in the social reasoning behind it. The United States has, for instance, been at war with one entity or another for all but a handful of the years of its existence, and virtually all societies expend a huge proportion of their resources on warfare or on preparations for it. For better or for worse, warfare is usually seen as an institutionalized expression of individual aggression, and since Darwin's time there has been much discussion of both in evolutionary terms. Most such exegeses view these behaviors in an adaptive context, invoking rather simplistic models like the "dads and cads" model of male aggression. This model pits altruistically caring and parentally investing "dads" against sociopathically cheating and raping "cads," and suggests that in the "evolutionary game of life" (for which read reproductive success), the cad strategy will work more often than not. Such simple binary models could clearly benefit from some bell-curve thinking, for very obvious reasons. Not every father is a monster or a saint.

Some evolutionary psychologists have suggested that the form of aggression we humans employ is actually quite complex, and that "aggression" as a category actually includes both offensive aggression and defensive aggression—two rather different things. They further argue that the existence of both types of aggression was, and is, subject to long-term adaptation. One route to this conclusion passes through the observation that our close living relatives, the chimpanzees, are rather aggressive and sometimes murderous creatures. This supposedly demonstrates that aggressiveness is ancestral in our primate lineage, implying that "genes for" it have been embedded in us by natural selection. However, chimpanzees are only one of two closest living relatives we have. The other is the chimpanzees' sister species the bonobo—and the bonobo is about as anti-aggressive as a higher primate species gets.

Other rather simplistic approaches have been taken to explaining aggression. Decades ago, it was thought that an extra Y chromosome in the genome of males made affected individuals more aggressive. Researchers looked at the karyotypes of men (about one in a thousand males will have that extra chromosome) and found a higher average number of Y chromosomes in men in prison compared to the unincarcerated, an idea that most likely had its origins in the undoubted fact that in general males are more aggressive than females, who completely lack Y chromosomes. This kind of reductive thinking—the more Y chromosomes you have, the more aggressive you are—was very attractive to the popular press. But a comprehensive study done in Sweden in the 1990s demonstrated that XYY males differed from other males in being taller than XY and XXY males. In all likelihood, this size difference was more closely correlated with the behavior in question than the chromosome was, and the researchers concluded that the extra chromosome contributed little if anything to an aggressive personality in human males.

Quantitative geneticists have also sought out specific "genes for" aggression and warfare. Such studies have included both candidate gene approaches (like Dean Hamer's "God gene"), and the now-familiar Genome Wide Association Studies (GWAS). A Finnish investigation demonstrated lower levels of both monoamine oxidase A (MAOA) and a neuronal membrane adhesion protein (CDH13) in violent criminals. One of these should sound familiar, as it is related to the biochemistry of monoamines, the very same small molecules that VMAT2 (the God gene) regulates. Monoamines are important in the brain's turnover of dopamine, a chemical that has great influence over our behaviors. CDH13 is also involved in regulating brain chemistry, so it is no surprise that it might be involved in aggression. In fact, in general the most promising candidate genes for regulating most kinds of up-and-down behaviors are those involving dopamine or serotonin. But here is where the half full–half empty problem comes in. Only 5 to 10 percent of the violent crime in Finland could actually be attributed to the variants of the two genes fingered.

To the researchers who undertook the study, it may have seemed that they had discovered a major explanatory component of aggressive behavior; but to most of us, if there were only a one in twenty chance of something being right, we wouldn't place any money on it. GWAS hasn't got us much farther, and as of 2018 the approach has been relatively unsuccessful in associating genetic loci with aggression and antisocial behaviors. Two loci, C1QTNF7 and DYRK1A, have been touted as showing antisocial associations, but in reality any functional role the former may have in aggressive behavior is entirely unknown, and all we actually know about the latter is that it is involved in abnormal brain development. Given all this, it is more than likely that what we define as aggressive or antisocial behaviors are much too diffuse to allow associating specific genes with them. On the other hand, the converse is almost assuredly true: that the genetic architectures of the behaviors are so complex that the association of specific genes can only explain a very small part of the variation present.

The ultimate act of aggression is killing. While individual murder has its obvious social taboos, and virtually every society condemns it, killing on a large scale in the name of warfare has been routine and acceptable among societies as far back as the record goes. How do we reconcile these conflicting attitudes? Malcom Potts and Thomas Hayden, authors of *Sex and War: How Biology Explains Warfare and Terrorism and Offers a Path to a Safer World*, have proposed that warfare is an evolutionary "hangover" related to adaptation to past conditions. Potts and Hayden essentially co-opt the "dads and cads" model to argue that organized aggression was adaptive in the human past, and simply blew up into modern warfare. Another evolutionary psychologist, Anthony C. Lopez, put it this way: "These adaptationist questions are necessary because, absent methodological innovations, there are only so many ways to interpret existing evidence of ancestral environments." In other words, Lopez is saying that because we can't explain something any other way, it must be due to natural selection and adaptation. This is a very familiar rationale in evolutionary psychology, and it rings exceedingly hollow in terms of everything we (and Lewontin and Gould) have had to say about chance and exaptation in evolution.

In a world in which both people and societies are in constant competition for space and resources, energetic interactions among entities are constantly and inevitably happening and must necessarily be resolved in some manner. Human beings are notably prosocial and cooperative, but high aggression certainly lies at one end of their behavioral bell curve, just as complete passivity does at the other. The world has had its share of Genghis Khans, just as it has had of its Mahatma Gandhis. As it happens, the human mean between meekness and mayhem may actually lie closer to the peaceable side than those of some of our close relatives; it has often been noted that if one were to pack three hundred chimpanzees into a Boeing 777 and fly them from New York to Tokyo, few would be left alive when the aircraft arrived in Japan. Humans may grumble under similar circumstances, but they typically accept the discomforts and indignities of long air trips with meek passivity. But then again, in sharp contrast to chimpanzees, our equally close relatives the bonobos tend to redirect social tensions in the direction of sex rather than violence, so that if we were to fly a plane full of them to Tokyo we would probably have an epic orgy on our hands. That's something it's hard to imagine we will ever discover for sure. But what we do know is that, no matter how docilely most of us behave on aircraft, there will be several murders in Chicago this week, and that it is unlikely that the United States will extricate itself from its endless succession of overseas wars anytime soon. That is because *Homo sapiens* has clearly retained violence as an option in both its individual and corporate behavioral repertoires; and as the world becomes increasingly crowded it is an option that, in the absence of appropriate social controls, is likely to be increasingly exercised. That does not mean, of course, that the genes did it. Once again, they are likely responsible for at most a tiny fraction of the variance in violent behaviors, most of which are intensely situational. If anyone ever tells you that our species *Homo sapiens* is "hard-wired" for any complex behavior, run a mile.

POLITICS

Many animal societies are hierarchical and highly structured, which is a characteristic they share with our politics. Nonetheless, because of the ways in which the human mind thinks and human political systems are organized, useful comparisons between animal hierarchies and human hierarchies are limited. So, although our colleague Frans de Waal made quite a splash with his book *Chimpanzee Politics*, and the relationships among chimpanzees and the coalitions they form are very intricate indeed, we will confine our attention here to human political behaviors, which have actually garnered quite a lot of attention from neurobiologists. There have also been some GWAS (association) studies of political behaviors, and some candidate (no pun intended) genes have been proposed.

A lot of evolutionary psychological attention to human political activity has been focused on the biology of leadership and what is called "followership," the idea being that humans possess specialized psychological mechanisms for approaching and solving problems of coordination. These mechanisms have, of course, been under intense selection for millions of years, and have been adaptive across the enormous range of human social structure and political systems. And beyond: as Mark van Vugt and Richard Ronay claimed, "leadership is abundant in the social animal world, from ants to baboons and from honeybees to humans." Once more we are in the realm of sociobiology; and once more we think that those authors are protesting a bit too much. In a general sense we have certainly inherited our intense sociality from highly social primate ancestors, but we doubt that the complexities of human political organization are direct translations of the dominance hierarchies of other higher primates—or indeed, that the qualities that make a modern human successful are directly comparable with those that make a successful chimpanzee. Clearly, "leadership" is a peculiarly human concept that implies something more than the bullying of dominant chimps. Evidently recognizing this, Herbert Gintis, Carel van Schaik, and Christopher Boehm suggested that "the heightened social value of nonauthoritarian leadership entailed enhanced biological fitness for such leadership traits as linguistic facility, ability to form and influence coalitions, and, indeed, hypercognition in general." But that may have been going a bit too far the other way, imputing a little too much to leadership and followership as guides to our diverse—and sometimes perverse—political systems.

Modeling can help shed light on the evolution of leadership, as witnessed by the work of the behavioral economists Ernst Fehr and Simon Gächter. These investigators used a standard public goods game, played by several subjects, to assess the effects upon groups of forced altruistic behavior. Under the rules of the game the general good of the playing group could be assessed, and free riders were able to take advantage of public goods to enrich themselves. In addition, those free riders could be punished or left unpunished, either by all players or by a predetermined leader. The results of this widely cited study are shown in the figure. As summarized by Mark van Vugt, "the striking result [was] that a single punisher was able to maintain almost the same level of cooperation as everyone punishing." Which suggested, of course, that the more parsimonious and less chaotic system involving a single leader was more effective than no punishment, and just as effective as multiple punishment.

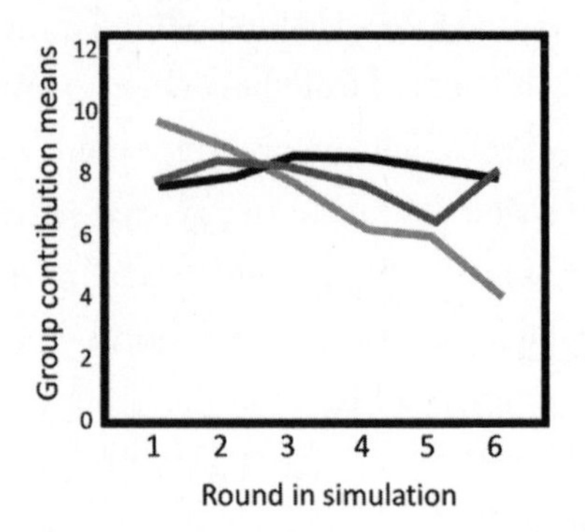

Results of Ernst Fehr and Simon Gächter's simulations showing the group benefits from the "all punish" experiment (medium gray line), the "leader punishes" experiment (black line), and the "no punishment" experiment (light gray line). *Adapted from King, et al. (2009).*

All of this makes great sense when we think about the inherently chaotic nature of human social groups; but van Vugt takes things a step farther to suggest that by "understanding the underlying biology of leadership we can better choose our leaders." We wish him luck with that, because the immense variation that exists among humans in responding to social situations, political or not, would almost certainly undermine any attempt to determine good leadership versus bad. This is where the bell curve intrudes, overpowering any knowledge we can infuse into rational political decision-making and leadership choice.

Amazingly enough, the heritabilities of various aspects of political preference have been analyzed using twin studies. They range from 0.6 (60 percent of the variance being genetic) for political knowledge to less than 0.1 for political party affiliation. Overall, though, the heritabilities of political "traits" (ranging from social trust, religious attitudes, and political donations to economic beliefs and attitudes to punishment) hover around 0.5. This looks relatively high; but it is nonetheless true that the precise genetic architecture of any of these traits is going to be exceedingly difficult to pin down, and we would hazard confidently that any GWAS study would find only a small proportion of the variance explained by associated genes.

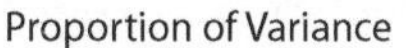

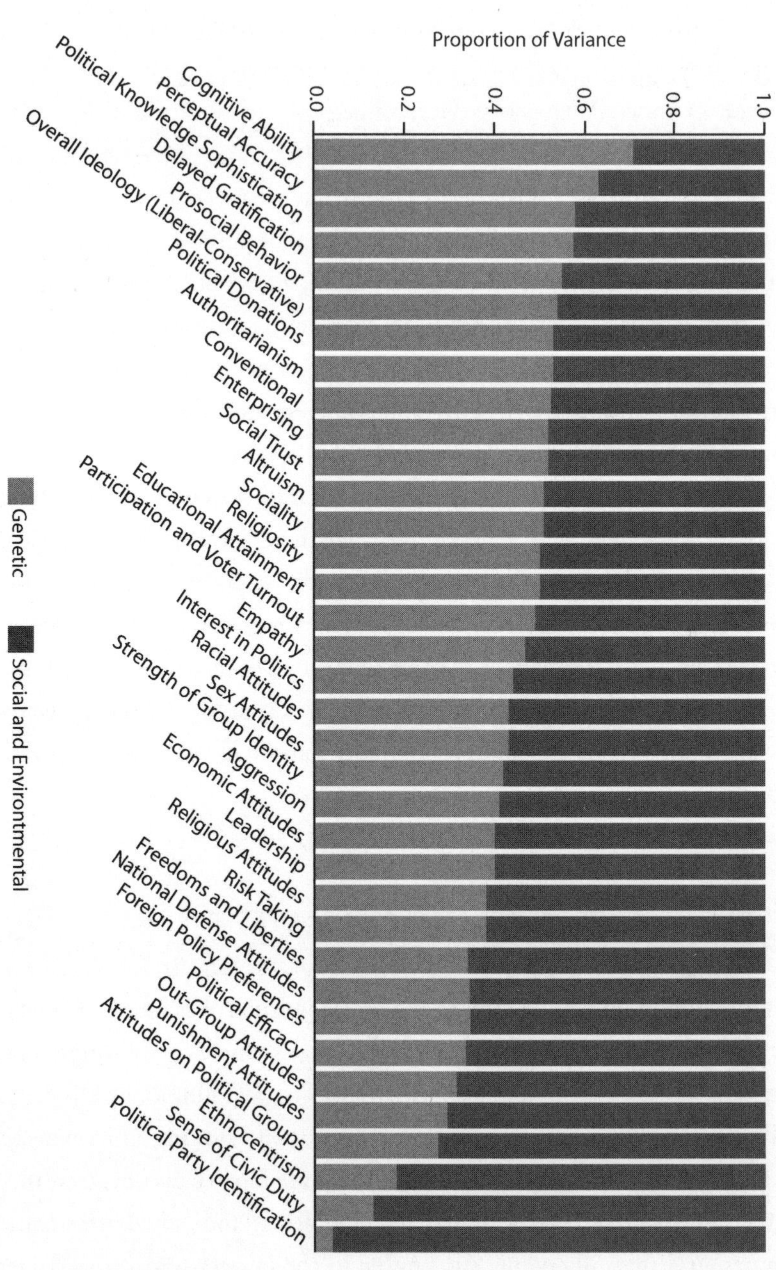

Graph showing heritabilities in some psychological and political traits, all characters of complex derivation. *Adapted after Lockyer and Hatemi (2018), graphic by Kayla Younkin.*

As with most of the behavioral traits we have discussed so far, two approaches have typically been taken to understanding the genetic architecture of political behaviors: the candidate-gene method and GWAS. Both approaches have shown some associations but, again, both suffer the same problems found previously. In a large study of the political attitudes of Australian citizens, Peter Hatemi and colleagues found a slew of genes that are involved in neural processing (serotonin, glutamate, dopamine, and so forth) of beliefs associated with both liberal and conservative political attitudes; but these associations could not be replicated in subsequent studies. In addition, no single marker region could account for more than twelve percent of the variance in the population. It is probably significant that the authors used a fifty-item questionnaire; and although that is certainly a lot of items (see our discussion of facial characteristics and landmarks), we highly doubt that such complex behaviors as political beliefs could be characterized fully with such an approach. Once again, while there may well be a genetic component to some or even all of the traits covered in the questionnaire, we suggest that they would most appropriately be understood as distributed along a bell curve. In politics above all, we can hope that we can find a majority in the middle ground.

SEXUAL BEHAVIORS AND IDENTITY

Reproduction is fundamental to life on this planet. Without it, life would screech to a halt. One reproductive strategy is employed by single-celled bacteria, which simply duplicate themselves by doubling their DNA (usually one single circular chromosome) and then budding off a new cell. There is a huge danger here, though, the same one we encountered in the case of the beloved Cavendish banana in which all individuals are identical clones, and thus all equally susceptible to the ravages of disease. Bacteria have gotten around this problem in a number of ways that are collectively referred to as horizontal gene transfer. Viruses may be recruited to shift

DNA between individuals, or the same thing may be accomplished directly, through physical contact. Either way, the genomes involved acquire variation. The eukaryotic organisms, among which humans belong, achieve the same result through sex.

Sexual reproduction in eukaryotes involves the fusion of gametes, sex cells derived from each of two parents, male and female. Each gamete contains one chromosome of each pair, so the resulting new individual has a complete chromosomal complement. Thanks to this diverse origin, and to the process of chromosomal recombination that takes place as the gametes fuse, each new individual will have a unique genomic makeup. There remains only the problem of how to get the gametes together, and the many ways nature has devised to do this are gathered by biologists under the rubric of sexual behavior. One approach, taken by eutherian mammals such as us, is for the male to fertilize the female internally, using a specialized organ. Clearly, for this to happen males and females of the same species must feel some innate urge to come together physically; and in intensely social higher primates such as we are, urges of this kind often blend in with a more general social bonding.

Since sexual behavior is so fundamental to the success of any species, the urge to mate has to be a very powerful one, and across the mammals the fact of possession of specialized sexual organs may express itself in a variety of ways. One variant is homosexual behavior, wherein the behavioral urge is directed toward members of one's own sex. Homosexual behaviors have now been documented in nearly three hundred different species of mammal, and it turns out that they are much more widespread in nature than this. Indeed, male fruit flies of the species *Drosophila melanogaster* may possess a variant of the fruitless (fru) gene that is associated with a tendency to copulate with other males. Similarly, another gene, genderblind (GB), which alters the processing of odorants (pheromones) in the brain, is also associated with bisexual mating behaviors among fruit flies. There is no equivalent to fru in the human genome, although if you search a human genomic database for GB you will get a "hit." This finding attracted a lot of interest from the press several years ago; but in fact, the hit pertains to a whole family of proteins, and not necessarily to the fruit fly GB gene itself.

Michael Bailey and colleagues have conducted twin studies of human homosexual behavior in both males and females. The overall heritability estimates for both male and female cohorts varied from 0.3 to 0.7, depending on the method used to compute them. That level of heritability implies a significant genetic component in this behavior. So once more enter Dean Hamer, who made a big splash with a "gay gene" even before he discovered the "God gene." Together with colleagues at the National Institutes of Health, Hamer reported in 1993 that he had associated male homosexuality to the X chromosome, and specifically to a rather large chunk of it known as Xq28. Over twenty-five genes are known to exist in this part of the human genome, which has been intensively studied because rare duplications in this area produce a syndrome characterized, among other things, by language development glitches, heart problems, and seizures. A study carried out six years later was unable to replicate Hamer and colleagues' results, but a study twelve years after the original again recovered a signal for male homosexuality in the Xq28 region.

As with the "God gene," there may be a genetic factor somewhere in this region; but in light of the group's overall results it is most likely of very small effect. Indeed, to the best of our knowledge every attempt to find a "gay gene," meaning a gene of major genetic effect in humans, has failed miserably. Attempts at GWAS of sexual orientation have recently focused on cases of gender dysphoria that led to surgical alteration of gender identity. For example, John Thiessen and colleagues did GWAS on a small cohort of transgender people, identifying twenty genes involved in gender identity. The core functions of these genes are centered in neurologic and hormonal pathways. All in all, the best guess is that gender identity is very complex genetically, and that it is probably determined via the interaction of hundreds of genes. This is very much what one would guess from the variety of gender identities that flourishes nowadays. In fact, gender identity in *Homo sapiens* seem to present us with a pretty good spectrum that runs the gamut of possibilities, and in which most of us, as usual, lie somewhere in the middle. Sexual orientation, which is not the same concept as gender identity, has a similar breadth of possibilities.

Homosexuality is only one of our human sexual behaviors for which scientists have sought a genetic basis. Others include sex appeal, male premature ejaculation, and marital fidelity or its opposite. These examples are hardly surprising, since the genetic factors that have been associated with these behaviors involve genes coding for serotonin, dopamine, and other brain reward pathways. Sexual dysfunctions seem to have been more intensively studied in males than in females, a skewed distribution that brings us to the sex-related behaviors that are grouped under the rubric of sexism.

The geneticist David Reich recently noted in a *New York Times* editorial that

> The differences between the sexes are far more profound than those that exist among human populations, reflecting more than 100 million years of evolution and adaptation. Males and females differ by huge tracts of genetic material—a Y chromosome that males have and that females don't, and a second X chromosome that females have, and males don't.

Here Reich summarizes the two major genomic aspects of male-female differences. The male Y chromosome is small (with only about sixty genes on it), and it is apparently shrinking. Still, it is not insignificant and, because males have only one Y chromosome, they are susceptible to genetic disorders related to genes on both the X and the Y chromosomes (men are both "X-challenged" because they do not have a backup X, and "Y-challenged" because they have only a single Y). The 100-million-year time period Reich mentions is the approximate amount of time the XY sex determination system has been around in our mammalian ancestors, and the impacts of sex differences in animals generally and in mammals specifically have been amply documented. Sexual dimorphisms (physical differences) abound in the animal world, resulting mainly from intense selection on traits related to reproduction, and sexual dominance systems can be observed in many groups of animals, especially mammals. The impact of those differences on social structures and on more specifically "sexist" male-dominant behaviors is not hard to discern, so that many researchers consider human sexism to be a passive behavioral correlate of the fundamental physical sexual differences dictated by the XY system. This would make sense, except that comparing sexist behaviors in our two closest relatives (bonobos and chimpanzees), suggests otherwise. This is because among bonobos females are dominant while, among the chimpanzees, males are. What's more, there are also mammal species out there in which sexual dimorphisms do not lead to dominance systems at all.

Attempts to discover the sociological basis of sexism led the psychologists Julian Oldmeadow and Barnaby Dixson to conduct a rather bizarre survey. They looked at a large sample of human males and tried to assess if facial hair was associated with sexist attitudes. The hypothesis was that men who do not shave deliberately want to emphasize the physical differences between themselves and females—which would indeed be a form of blatant sexism. There is even a term for not shaving: pogonotrophic behavior. Oldmeadow and Dixson's results indicated that pogonotrophism was indeed correlated with sexual attitudes: you are more likely to be sexist if you grow a beard. But again, we have to remember that correlations and causation are different things (as in the earlier example of the number of firefighters and the dollar amount of fire damage of a fire, which really wasn't a correlation, but the result of a chain of correlations). Pointing out that apparently nice correlations are often invalidated by omitted third variables, the lavishly bearded Kahl Hellmer and colleagues studied a wider range of behavioral features, finding no direct correlation between facial hair and hostile sexism. Instead, they implicated low empathic concern as the primary driver of sexism of this kind. And low empathic concern doubtless has its own causes and correlations. So once again we are faced with a complex behavior with complex causes and can only emphasize, along with Reich, that "we should both recognize that genetic differences between males and females exist, and should accord each sex the same freedoms and opportunities regardless of those differences."

ECONOMICS

With complex societies came complex economies, and with complex economies came a whole new kind of decision-making. Economists have long made the rosy assumption that markets are rational, but over recent years it has become clear that they are driven by millions of individual decisions that may not be rational in the least. We pride ourselves on our rational faculties, but in fact we are notoriously bad judges of such things as risk, for example fearing flying while not giving a second thought to the much more dangerous car journey to the airport.

A whole new branch of economics—neuroeconomics—has sprung up to study the tendencies people show in their decision-making processes. One very clear tendency that has emerged in numerous studies is that most people have a very strong aversion to loss, a fear that apparently typically trumps their desire for gain. The jury appears to be out right now on whether this fear is mediated by a particular part of the brain or results from a more general process of reasoning, but subjects in neuroeconomic experiments tend to show signs of higher stress when faced with the prospect of monetary loss than they do when anticipating equivalent gains. So, however it is generated, the experience of loss seems to be the stronger one. The take-home message here seems to be that nature has very definitely not skewed us in the direction of rational decision-making, at least in this domain.

One thing that neuroeconomics seems to have shown pretty clearly is that there is frequently conflict between the kinds of decisions we make using the more ancient parts of our brain (for example, the limbic system, discussed earlier, which includes such structures as the amygdala and hippocampus, and mediates emotions and memories), and using the more deliberative newer parts such as the prefrontal cortex, which makes rational decisions. If you have to choose between an immediate reward now or a greater reward tomorrow, for example, you will have to resolve a conflict between the limbic system, which will tell you to grab the proffered reward, and the frontal cortex, which will tell you to wait. This kind of conflict is as much the rule as the exception whenever you call on your brain to make a decision, and it makes the organ a pretty typical product of a rather messy evolutionary process in which all innovations have to be tacked onto, or at least be compatible with, whatever structures preceded them. Conflict and conflicted behaviors are far from unique to humans, but in our case the mental processes unleashed by conflict are particularly complex, and they do not necessarily result in optimum resolutions.

Although economics is often called "the dismal science," economists usually make the rather optimistic assumption that, at least in their realm, humans are rational actors. So how about a genetic basis for our vaunted rationality? It is obvious from the outset that any answer to this question will be very complex indeed, but first we need to determine exactly what is involved in our economic behaviors. Economists love playing games, so to do this let's play a game we call Paul Glimcher Took My Wallet. The eponymous Paul Glimcher is one of the architects of neuroeconomics, and when he came to the American Museum to give a talk he played the following game with us. In response to a question about rationality, he asked us to envision a fruit market in which consumers preferred apples over pears, pears over bananas, and apples over bananas. This can be symbolized as A > P, P > B, and A > B, in a straightforward logic system. The object of the game is to start with an unfavored fruit and to end up with the fruit you like most. If you start out with a fruit you aren't terribly enamored with, you can trade for a preferred fruit if you pay one dollar. As sellers, we engaged Glimcher in a transaction in which he had started with a pear that he wanted to trade for an apple. Transaction over, we both went away happy. We earned a dollar and got a banana to sell later, while he got his apple for only a dollar. Next, we were the consumer. But this time an irrational choice system was established for us. In this system we would prefer bananas over pears, pears over apples, and apples over bananas, or B > P, P > A and A > B. Glimcher started us out with a banana. Our logic system told us that we liked apples better, so we gave him a dollar and traded for the apple. But wait, he also had pears. Since our logic system tells us we like them better than apples, we traded him for the pear, and paid another dollar. So we were getting what we wanted, but it was getting a little expensive. However, we realized then that we liked bananas better than pears, so we had to pony up again to get one of those. We were now three bucks in the hole, and we were holding the very item we had started out with. At this stage in the game we began to get the point, and we dropped out while Glimcher tried to hide the sly grin on his face.

This embarrassing episode taught us clearly that any economic system must be based both on a logic (there have to be rules) and on rationality (the rules should be rational). Variations on this theme are, of course, how one makes money by manipulating the logic and rationality to maximize gains and minimize spending. And while this is a very simplistic example, it does show the basic needs of any economic system.

In the Ultimatum game we mentioned earlier the logical thing to do is to play forever, although most players will terminate the proceedings when their sense of fairness overcomes rational calculation. But players with certain neurological conditions such as autism will play the game more rationally. An abundant literature shows that autism has a genetic basis, but that basis has not yet been nailed down, for two reasons. The first reason is simple: namely that autism is complex. The second concerns the difficulty of characterizing any complex system, because autism is a broad-spectrum phenomenon (meaning it comes in many kinds that have so far confounded genetic dissection). If a behavior like autism can be expressed in logical responses, it means that *any* rational behavior that incorporates logic will start out as being very complex. If we then add in the genetics, and the interaction between logical and rational behaviors, we will complicate the situation even further. Once again, we are faced with an astonishing complexity of causation at the genetic level: one that inevitably, as Fisher showed, resolves down to a normal distribution.

INTELLIGENCE

Intelligence is probably the most visible, important, and hotly contested complex behavior that humans exhibit. And while other animals can clearly show high levels of what we might call intelligence, and this quality clearly comes in many forms, it remains uncontestable that humans are unique in the ways in which they use information and in how they employ their brains to cope with their environments. What is more, there is no question that humans are highly variable for the trait of "intelligence"—which has, indeed, been made the poster child of the bell curve, albeit with dubious results. A full review of intelligence would fill several scholarly volumes (and has done so, along with some not-so-scholarly ones), so here we will just focus on one interesting study that appeared just as our manuscript was due to our publisher. This study used millions of subjects' genomes (partially provided by the commercial genomics outfit 23andme) to examine genomic associations with educational attainment. We are aware that educational achievement is not necessarily a good direct proxy for what we might want to call intelligence, but it does nevertheless tell us something about the architecture of intelligence as humans develop. And just as notable as this study's scholarly content is the response the it has received from educators and journalists.

The study revealed that a whopping 1,271 genes are involved in a mere 10 percent of the variation in educational attainment. As we have seen with many other complex traits and their genetic architectures, this is a lot of genes to explain rather little of the variance in any behavioral capacity. This by itself suggests an inherent complexity in the trait involved, and it gives us some idea of the difficulties we face when merely trying to define "intelligence." Yet, well-respected educators and others have already suggested that this study is a significant contribution to educational understanding. In a *New York Times* editorial Kathryn Paige Harden, a psychologist whose specialty is adolescent development, suggests that there are two ways we can use this genetic "information." The first is to point out that success in our educational system is not entirely reliant on merit. As she says, we don't earn our genes; and because the inheritance of our genes is somewhat random, then "everyone should share in our national prosperity, regardless of which genetic variants he or she happens to inherit." Second, she suggests that knowing the genetic architecture of educational attainment will help scientists identify the environmental effects that interact with the genes that influence educational attainment.

We suggest that, while Dr. Harden's intentions are admirable, there are some basic misunderstandings about genetic architecture here. Dr. Harden is at pains to dispel eugenic intentions in her editorial; but alas, the specter of this eight-hundred-pound gorilla in the room can never be eliminated when the matter of the genes is raised in any discussion of intelligence. And intelligence is far too slippery a concept to be discussed usefully in the context of genes, consisting as it does of numerous different forms of acumen, each of which doubtless has a very complex genetic underpinning. What is more, all those capacities are distributed across the population in a noncorrelated way. We thus strongly doubt that variations in performance in the many different kinds of tests available will be usefully addressed by associating yet more genes. Instead of a reductionist approach to traits like intelligence and educational attainment, we would be better off trying to assess the whole educational spectrum and its characteristics in the wholly desirable attempt to raise general educational attainment. Rather than particularizing, we should recall that in the social realm, as in the biological one, the only available target of selection is the whole organism, not its individual characteristics.

Our scamper through a miscellaneous selection of human behaviors shows pretty clearly not only that attempts to associate human genes with particular complex behaviors have yet to get us very far, but that they are rather unlikely to get us much farther in the immediate future. Human beings are extraordinarily complex creatures, and they interact with their environments as whole individuals, rather than as carriers of particular behaviors. Of course, even a single behavioral act might conceivably seal the future reproductive fate of an individual; but this will be the exception rather than the rule. And in the long evolutionary run the fate of the species will count immeasurably more than that of the individual—or his or her features, or the genes that explain only a tiny fraction of the variance in them.

It goes almost without saying that our biology, encoded in our DNA, certainly counts; and it daily constrains what we can do. And we are equally clearly the product of a long evolutionary history that in a very real way accounts for what we have become. As a result, we *can* in a broad sense blame our biological heritage for our slipped disks, or our wrenched knees, or our gastric reflux. But railing against history will not help much here, and it won't help a lot more when we contemplate our behavioral imperfections. As individuals, we human beings are commonly very much the creatures that our DNA specifies, although to the extent that our rational capacities can override our emotional urges, we do have choices in the ways in which we behave. But as a species, all bets are off. Are human beings heterosexual? Yes and no. Are they greedy? Yes and no. Do they believe in a god? Yes and no. Do they commit infidelity? Yes and no. You get the point.

FIVE

HUMAN BEINGS AND CHOICE

Human beings and their progenitors have traversed an amazing amount of biological and behavioral territory in the last several million years. Starting off as relatively routine if rather oddly proportioned apes, members of our lineage rapidly transformed in myriad ways, cognitive and physical alike. Indeed, so great was the metamorphosis that the distinguished German scientist Bernhard Rensch once proposed that *Homo sapiens* should be classified in its own kingdom, Psychozoa, a reclassification that would place our single recently evolved species on a par with the major groups containing *all* of the plants, fungi, (other) animals, and so forth.

This suggestion clearly went way over the top in forgetting how closely we are bound by ancestry to the rest of the living world, but the fact that it could be made at all points to just how strange an animal we are. And how unprecedented: there is little in the older fossil and archaeological records that would predict anything about what we would eventually become, certainly in the cognitive sense. In turn, this points to the fact that we were not groomed by nature over the eons to assume our unusual role. Instead, the cognitive uniqueness that provoked Rensch's proposition was acquired rather recently, in a relatively short burst of innovation that climaxed in an abrupt qualitative change in the way we process information. There is a great deal we don't know about exactly what happened; but, whatever the details, we can confidently say that as a result no creature living in the world today is more unlike its ancestor of two or three million years ago than we are—no matter how you measure that difference.

So there really is no question that, particularly since the emergence of our own genus *Homo*, hominids have evolved astonishingly fast. But whether this fact holds any lessons for the human future is another question entirely. Sure, a look back at our history of dramatic evolutionary transformation makes it sorely tempting to extrapolate this powerful pattern of rapid change indefinitely into the future—all the more because the prospect of future biological improvement offers a ray of hope for a species that has, time and again, let its individual members down in hideously traumatic ways. For, while the history of humankind has had its dazzlingly brilliant high points, on the other side of the ledger it is littered with disreputable-to-disgraceful and often violent episodes that one shudders even to contemplate. It is those historical nadirs that above all make it hugely appealing to imagine that in the short period covered by written records, we have been witnessing merely a passing phase: a phase that lies on an evolutionary trajectory of change that must surely be carrying us irresistibly toward a state of greater rationality and responsibility.

If only this were true! But alas, the probabilities lie elsewhere. Of course, in making any kind of prediction about the human future we are handicapped by our ignorance of just what it was that drove the hominid lineage toward an unprecedentedly high rate of evolutionary change. We can speculate that our extremely rapid evolution was somehow related to humans' unusually intense sociality, and to the interplay between the consequent complexity of human intragroup relationships and our developing technological prowess. But while all those factors were clearly formative in what we have become, just how they might have affected the rate of change in hominid evolution, and what other elements might also, or alternatively, have been in play, remains impossible to say. For the time being, then, speculation is currently the only weapon we have in our locker for resolving the question of where our behavioral repertoire is headed; and we will consequently have to look to alternative lines of evidence if we remain rash enough to try predicting our biological prospects.

LOOKING BACK

So, let's start with one thing we really do know for sure, namely that our genus *Homo* evolved during a time of extraordinarily unsettled environmental conditions. This is because for much of the period over which mankind's precursors evolved, our planet was in the grip of what are informally known as the Ice Ages, a geological interval during which world temperatures fluctuated between periods of extreme cold and more temperate times such as those we are experiencing now. Latterly, there has been a fairly regular one-hundred-thousand-year swing between peaks of warm and cold, though there were many minor climatic oscillations in between those climatic apexes and nadirs.

All of those fluctuations in temperature and humidity across the globe translated directly into dramatic variation in the environments in which our ancestors lived. Sometimes such variation was amazingly rapid, occasionally occurring on the scale of a single individual's lifetime. And hardy and adaptable as early humans undoubtedly were, they could never have simply shrugged off the resulting stresses; after all, their very lives were put at risk by climatic oscillations that wrought repeated changes in the floras and faunas on which they depended. The demographic results of those changes would thus have been dramatic. As secondary carnivores—creatures that had only lately come to depend on hunting as a way of life—early populations of *Homo*, unlike their predecessors, had never been particularly common on the landscape, not least because predators are of necessity considerably rarer than their prey. But in good times animal and plant resources would have been abundant in the environment and would have supported relatively larger and more continuous hominid populations, while in leaner periods hominid numbers would have contracted strongly, thinning out already sparse local populations. Local extinctions as conditions grew more difficult would then have assured the isolation of surviving populations.

As it happens, periods of constant population expansion and contraction provide ideal conditions for the accumulation of genetic changes. This is because, for the same reasons of chance that we looked at earlier, randomly occurring evolutionary novelties that could never make a dent in a large population have a chance of becoming "fixed" in the small ones that population fragmentation produces. Conversely, population expansions in good times lead to the competition, mingling, and displacement that are the lifeblood of the evolutionary dynamic. Of course, the fact that accelerated evolution is not seen among the hominids' ecological competitors suggests that environmental change can never account for everything—and indeed, it implies on its own that it was the hominids' response that was special, rather than the stimulus itself. But it is nonetheless evident that the Ice Ages provided the ideal background conditions for rapid hominid evolution. And, whatever the exact reasons, the archaeological record clearly shows that the hominids duly responded in spades.

Contrast those past demographic conditions with the situation today. From early times hominids had been hunters and gatherers, living off nature's bounty as it presented itself to them. They were regularly on the move, following their migratory prey, or traveling to the places where ephemerally fruiting trees or other seasonal plant foods were to be found. This strategy would have been variably expressed from one locality and latitude to another, with greater or lesser dependence on hunting or on gathering, according to the resources available. But, however their precise economic strategies might have varied over time, and however intensive their exploitation of available resources might have been in any particular place, the lifestyles of our ancient ancestors were itinerant ones. And accordingly, the number of individuals in any local population was controlled largely by the resources available. This was true even of the earliest symbolic *Homo sapiens*, whose cognitive processes were fundamentally the same as ours, but whose subsistence strategies were profoundly different from those practiced by the vast majority of people living now. Those early moderns continued to live entirely by the rhythms of nature; and although their new cognitive skills allowed them to exploit their habitats more intensively than their predecessors had done, and thus to exist at higher densities, by today's standards they were still extremely thinly spread across the landscape.

But then something extraordinary occurred. As the last Ice Age began to wane some twelve thousand years ago, climatic warming led to the contraction of the polar ice caps and to major environmental change throughout the planet. In itself, this was nothing new. But the human response to the changing conditions was, because those early symbolic humans reacted to their changing habitats in an entirely unprecedented manner. Instead of allowing their populations to adjust in the interest of preserving their traditional ways of life as their ancestors had done, people began to adopt a radically new economic strategy. They started to settle down, to cultivate crops, and to domesticate animals.

As you might expect for a species that had been cognitively modern for some time, not all of those activities were entirely unanticipated in earlier ways of life. It is believed, for example, that some hunters and gatherers in Europe may have controlled reindeer herds in Lapp-like fashion well before the end of the last Ice Age. There is evidence in the Near East for a dietary shift toward cereals even before that Ice Age peaked, and a group of researchers working in Jordan has recently reported that hunter-gatherers there were making a sort of bread as long as 14,400 years ago, also before the warming set in. And earlier yet, hunters living on the Central European Plain some 15,000 years ago had already contrived to prolong their local sojourns (in some very impressive huts built out of mammoth bones) by storing reindeer meat in freezer pits dug in the permafrost. So, something behaviorally unusual was clearly already in the air before the ice caps began to retreat.

Nonetheless, total commitment to the radically new settled way of life was entirely without precedent; and, once it was embarked upon, in some places the conversion to the new economic model proceeded extraordinarily fast. For example, between 11,500 and 11,000 years ago a site in Syria called Abu Hureyra was just a place where hunters and gatherers temporarily camped. By 10,400 years ago residents of this same locality were supplementing their traditional diet with cultivated cereals, and by 9,000 years ago they were fully committed to sedentary life and to livestock and cereal domestication. A total economic transformation, the most profound in the history of humankind, was completed in a mere two millennia!

As if in deliberate testimony to a new economic spirit that was suddenly alive in the world, similar developments took place independently in several different areas around the planet. In addition to the Fertile Crescent of the Near East, settled lifeways and animal and plant domestication were adopted at various times between about twelve thousand and seven thousand years ago in Central and South America, New Guinea, China, India, and even on the fringes of what is today the Sahara. And once the dynamic of sedentary existence had established itself, it developed with similarly amazing rapidity. In Mesopotamia (a center of early civilization located in the area between the Tigris and Euphrates Rivers in what is now Iraq) the typical individual living eight thousand years ago would have belonged to a relatively egalitarian rural society of a few hundred people at most. All community members would have lived very similar lives, occupying similar dwellings and sharing broadly the same skill sets. In dramatic contrast, a mere three thousand years later the same individual's descendants might have been slaves, or laborers, or craftspeople, or merchants, or even royalty; and they would quite likely have lived in an economically complex city that harbored several tens of thousands of inhabitants living in anything from hovels to palaces.

Once the Faustian exchange of hunting and gathering for settled existence had been made, the die was clearly cast. Human beings were poised to step outside of nature and to consciously differentiate themselves from their surroundings. Early hunter-gatherers had most likely thought of themselves as part of the natural world: as integrated, in some profound way, into the biota around them. Certainly, to the best of our knowledge they behaved that way. But once farming had been adopted, nature became the enemy. For the rain has never fallen and the sun has never shone at the farmer's convenience. Once adopted, agricultural life quickly and inevitably became a battle with the elements: a struggle to overcome the vicissitudes of climate by the application of technology and sheer muscular effort. Inevitably, this struggle soon transmuted into a bid to dominate nature, spurred along by associated feelings of entitlement.

Our colleague Niles Eldredge accordingly suggests it is no coincidence that the founding texts of the Judeo-Christian religions—which are the closest thing we have to the direct testimony of the early farmers of the Fertile Crescent—contain what he calls "the most ringing Declaration of Independence ever set down." He is referring to the exhortation in Genesis 1:27, which translates as: "God said be fruitful, and multiply, and replenish the earth, and subdue it; and have dominion . . . over every living thing that moveth upon the earth." This is the doctrine of a creature that has ceased to feel part of the natural world and instead has come to see itself as opposed to nature. And once that point had been reached, it was probably inevitable not only that humans would eventually despoil the planet, but that someone would sooner or later think it appropriate to classify human beings as psychozoans, rather than as primates.

The feeling of apartness from nature that developed in the wake of settled existence is fundamental to the way human beings have subsequently done business in the world, bending the environment to their will by the exercise of ingenuity and sheer hard work. Indeed, it lies at the very foundation of who we think we are, and it is integral to our modern identity as human beings—which thus emerges as a recently acquired trait. The philosophical and psychological implications of the sedentary lifestyle are thus immense. But more important in practical terms are its economic—and more specifically, demographic—consequences. For, while hunter-gatherer population sizes typically fluctuated in response to changes in the natural abundance around them, the early manifestations of farming required the labor that only accelerated procreation could provide. Inevitably, this necessity provoked a spurt in human population growth that was abetted by practical circumstances.

Itinerant mothers, especially hunter-gatherer mothers without access to beasts of burden, are severely limited in the numbers of children they can transport with them across the landscape; and it is surely no accident that in historical times hunter-gatherer mothers in the Kalahari Desert breastfed for longer than any other people ever documented, delaying by years the onset of their first postpartum ovulation. In contrast, in a productive environment the number of children sedentary mothers can keep around is limited only by their reproductive capacities. Settled lifestyles everywhere have consequently been associated with demographic booms. Figuring out ancient population sizes on hugely incomplete evidence is always to enter on hazardous territory; but one mainstream estimate puts the world human population ten thousand years ago at a maximum of two million individuals, most of whom were still hunters and gatherers. From that point, the population curve moved exponentially upward, as agriculture and urbanization inexorably grew, and the original hunting-gathering way of life became steadily expunged from the human repertoire.

LOOKING FORWARD

Today the world population of *Homo sapiens* is well on its way to eight billion, and it's projected to surpass nine billion by the middle of this century.

Cultural accommodation has allowed human beings to colonize every continent on the globe, mostly in mind-boggling densities. At the same time, mass transportation has blossomed, facilitating an increasing globalization of cultures and peoples. This jam-packed and constantly churning world is, to put it mildly, a far cry from the sparsely populated planet of hunter-gatherer yore. And its implications for our original question concerning future human biological evolution is glaringly obvious: current demographic circumstances are entirely inimical to the fixation of significant spontaneous genetic novelties in the human gene pool, and thus to the prospects for future biological change in our species.

Nonetheless, it is common for biologists and others to claim that our species is still evolving in response to selective pressures. And yes, before you mention it, certain dairy-oriented populations do appear to have acquired high levels of lactose tolerance in the relatively recent past. But this is a rare example of what was evidently very strong selection during a time when populations were still relatively small; and although gene frequencies will doubtless continue to slosh around in regional subpopulations of our overcrowded human species, the chances that the human gene pool will contrive to incorporate any really meaningful genetic novelties in the future are basically slim to none.

What is more, true evolutionary change requires more than just those spontaneous novelties to play around with. It also requires speciation, the process whereby a parent species buds off a population that not only possesses its own genetic peculiarities (because not all subpopulations of the parent species will have been genetically uniform), but that somehow acquires reproductive incompatibility with its parent. Only by attaining this incompatibility through speciation will the new entity assure the permanence and historical independence of both itself and its unique characteristics. In the early stages of the speciation process partial compatibility may linger, so that if contact occurs between a parent and its daughter species, or between daughter species of the same parent, a degree of gene exchange will remain possible. This evidently happened in the case of *Homo neanderthalensis* and early *H. sapiens*, in which DNA evidence suggests limited intermingling. Still, once sufficient incompatibility has been acquired, the offspring species will finally be definitively free to follow its own independent evolutionary trajectory. That point may actually be reached before isolation is entirely complete, as we also see from the example of those late Pleistocene hominids. For, as we've noted, despite some apparent minor exchange of genes after modern humans had physically invaded its territory, *H. neanderthalensis* rapidly became extinct without any evident physical alteration, while a substantially unchanged *Homo sapiens* went on to dominate the globe.

Among mammals like us, the establishment of reproductive isolation requires prior physical isolation. There is simply no way we know of in which conspecific populations still in active contact can differentiate sufficiently. So, the lesson here is quite transparent. In our crowded and increasingly mobile human world, the chances of geographical isolation among any *Homo sapiens* populations are effectively zero, along with the prospects for speciation and for any divergent evolutionary courses in time to come.

The implications for the human biological future follow as night follows day. Under current circumstances, the probability of imminent significant biological change in the human lineage is negligible. But that is, of course, under current circumstances; and there are two conceivable exceptions that might promote biological innovation in our lineage. The first and most obvious possibility is that demographic change *will* indeed occur: something will happen to dramatically reduce the human population and to fragment the survivors into small isolated pockets, reestablishing the conditions required for evolution. Any number of scenarios lying well on this side of science fiction might fill the bill here. Worldwide nuclear conflict no longer seems the unthinkable thing it once was; a lethal, robust, and easily transmissible virus might easily emerge and ravage the densely packed world population; or an asteroid might hit our planet with consequences similar to those that carried away the nonavian dinosaurs some sixty-six million years ago. Nonetheless, given the huge human population that currently envelops every habitable surface of the globe, and with the abundant protective technologies available, an event in any of these categories would have to be extreme indeed to make the required demographic impact.

And even then, the subsequent future direction of human evolution would basically be impossible to predict simply from the extrapolation of past trends. To understand this, let's look at two major features of human evolution that futurologists have often seized on to predict what humans might become. One of those features is the unusually light construction of our bodies. Our bones are notably less robust than those of extinct close relatives such as *Homo heidelbergensis* and *H. neanderthalensis*, whom we can correspondingly conclude were a lot stronger and more muscular than we are today. By projecting the "gracilization" of *H. sapiens* into the future, and by combining the resulting body form with the extraordinary technologies that now substitute for much of the work that used to be done by our arms and legs, it has been predicted that humans will in the future become slenderer and weaker-limbed yet—in extreme visions, even losing the ability to move around without technological assistance.

Still, while this all sounds perfectly logical, the problem is that gracilization does not actually seem to have been a long-term trend in *Homo sapiens* evolution. Rather, it seems to have been a one-off event that occurred at the origin of our species. What is more, that event gave rise to a product that was self-evidently highly successful. And, as such, our current body form is something that is much more likely to be stabilized by future selection than it is to be changed by it.

Similarly, there is the famous trend (or rather, several parallel trends) toward human brain volume enlargement that has unfurled over the last two million years or so. During this time average hominid brain sizes have expanded by a factor of three, though the exact pattern of that expansion is not understood: what we perceive as a trend might, in fact, have been achieved by numerous step-wise shifts, separated in time, involving many species, and not all even necessarily in the same direction. Nonetheless, futurologists have taken the apparently inexorable overall hominid brain expansion to imply that, in years to come, the increasingly gracilized bodies of our descendants will support ever-enlarging heads that contain increasingly swollen masses of brain tissue. That extra tissue, it is believed, would make its possessors much smarter than we are today—whatever those additional smarts might actually translate to in practice. But while members of a species that prides itself on its brain power might jump to the conclusion that becoming "more intelligent" in this way would be an unalloyed good, in the real world everything comes at a price. For, as we have already mentioned, your brain is a very metabolically expensive organ. And that is a really good reason not to have more of it than you absolutely need.

We can't think of a better example to illustrate this fact than the 13 percent *reduction* that has taken place in average *Homo sapiens* brain volumes over the few tens of thousands of years that have elapsed since the late Pleistocene. As we explained earlier, we believe that this reduction is related to a greater energetic efficiency of the symbolic cognitive algorithm compared to the earlier intuitive one, so that once you have symbolic cognition, you no longer need a brain as big as a Neanderthal's. What is more, as we argue below, future cognitive refinement will almost certainly come from exploring a capacity we already have, rather than from any biological innovation. And with multiple reasons for believing that future brain enlargement is hardly in the cards, simple extrapolation of apparent past trends to predict future change really doesn't seem to fit the bill in this case, either. All in all, in the highly unlikely event that the *H. sapiens* lineage does undergo further significant change in either the physical or the cognitive domains, all bets are off on what that change might involve.

The other possibility often mooted for future change in the human population has to do with artificial genomic manipulation, often known as genetic engineering and more recently also as gene targeting, replacement, or editing. One popular misconception is that genetic engineering is something new. That's actually not the case, since Mother Nature herself has been doing it since the beginning of life on this planet, and animal and plant breeders have been hard at it since the early Neolithic. The main difference from the natural process is that, instead of occurring spontaneously and then being subject to natural selection, the changes in the DNA wrought by genetic engineers are specific, fast, and intentional. And of course, where there is intention there are invariably unintended (and unpredictable) consequences—a fear of which constantly haunts public attitudes toward the practice, along with the ethical concerns raised by almost any human intervention in natural processes.

Those fears have only increased with the accelerating speed and technological ease with which genes can be targeted and manipulated. Over the last four decades or so, numerous ways of doing this have been introduced, the latest of which is the gene-editing technique known as CRISPR-Cas9. Borrowing the necessary biochemistry from the immune systems of bacteria, biologists can readily recruit the cell's own repair mechanisms to edit genes or render them nonfunctional in an organism's genome by "cutting and pasting," as it were, from one organism to another (or even inserting DNA synthesized in the lab). What is more, such techniques can be combined with the insertion of "gene drives"—stretches of DNA that truly deserve to be called "selfish genes" since they contrive to appear in offspring in greater frequency than normal genes do, and that thus rapidly spread through the population. The first widely publicized practical application of a gene drive involves controlling populations of disease-transmitting mosquitoes through spreading sterility among population members. The jury is still out on how effective this gambit will prove in practice, but meanwhile researchers have already succeeded in the more difficult technical task of introducing a gene drive into a mammal, the laboratory mouse.

As its name suggests, CRISPR-Cas9 has two components: the enzyme Cas9 that can be programmed to cut out a specific piece of DNA, and the "guide RNA" that transfers the enzyme to its destination on the DNA molecule along with a replacement sequence. Researchers at the University of California, San Diego, recently announced that they had succeeded in modifying the genomes of female mice to carry the gene that specifies the desired Cas9, and those of males for the guide RNA. When the modified males and females bred, they produced pups that had both of the CRISPR-Cas9 elements, albeit on separate chromosomes.

Cells have very efficient repair mechanisms that mend broken DNA. Preferentially, they simply join broken DNA ends together, but an alternative mechanism inserts a new sequence of DNA into the gap. The mouse researchers took advantage of this alternative to put in a copy of a modified coat-color gene—a ruse that succeeded in females but, for technical reasons, not in males. No worries; in the females, the new coat color gene was copied to the partner chromosome in as many as 79 percent of the eggs produced. If the female concerned was then mated to an unmodified male, the odds that offspring would inherit the modification rose to around 90 percent. This is still a lot less efficient than gene drives are in mosquitoes, and the inconvenient fact that CRISPR and Cas9 are carried on different chromosomes in the laboratory mouse means that with the passage of generations they will inevitably become disassociated. Nonetheless, the experiment still qualifies as a proof of concept in organisms as complex as mammals; and if such techniques are possible in other mammals, they will sooner rather than later be applied to humans.

The introduction of new genes—especially rapidly spreading ones—into the human genome is a prospect that excites some and alarms others. At the very least it raises a host of ethical questions, so it may be reassuring to reflect that gene drives work best in quickly reproducing organisms, and that humans reproduce very slowly. Oddly, researchers have begun to use CRISPR-Cas9 to edit the genome of another slow reproducer, the elephant, using woolly mammoth genes as templates. The end purpose of this effort is to produce a woolly mammoth from an elephant genome. As of January 2018, about fifty of the tens of thousands of elephant genes have been edited to resemble woolly mammoth genes. Such research has been justified as a potential pathway for reviving our planet's "necrofauna"; in other words, as a way of raising the dead via bringing back vanished species from extinction.

If gene editing is possible in other mammals, it will sooner rather than later become practicable in our species. The most immediate application of gene-editing techniques among humans will be therapeutic, with stretches of normal DNA substituted for mutated ones that cause disease. But gene editing's potential for producing "designer babies" is obvious; and in China CRISPR-Cas9 has already been employed to determine whether it is possible in principle to genetically modify nonviable human embryos, albeit with mixed results. All in all, on the technological level the rather limited approach of embryo selection (producing multiple embryos from the same parents, analyzing their DNA for defects and other features, and choosing among them) seems for the immediate future to be the most promising avenue for obtaining the kind of offspring that parents desire. What's more, while in conjunction with cultural biases deliberate selection of this kind by individuals may produce quite severe social problems—just for a start, think of the current acute shortage of brides for young males in China—there is no immediate prospect of material change that will affect the species as a whole. Both in nature and in the laboratory, new forms have always been produced by the isolation of strains within species; and under current demographic—and ethical—circumstances, such isolation within our species *Homo sapiens* is pretty unimaginable.

Finally, quite apart from the major ethical problems human gene editing creates, especially in light of our discussion of the complexity of most traits, particular problems also arise when we think about using it to alter human behavior. While researchers can easily do trial experiments using mice, flies, and plants, the trial phase on behavioral traits in humans (intelligence is the complex behavior most often mentioned in connection with gene editing) could be terrible and cruel. When only a small percentage of the heritable aspect of most complex behaviors is explained by genetic variation, tinkering with this incomplete amount of the genome would be a hit-or-miss process—something not allowable when we are talking about human life. We also do not know how much of many behaviors is what we call "epi." Our epi problems are twofold: epistatic and epigenetic. Epistasis is basically the idea that a single gene can—and most likely will—have many interactions with other genes, as we discussed earlier; and because most of the genes in our genome have multiple roles in genetic pathways and cascades, editing it to enhance one pathway might cause it to misfire in another. Our second "epi" is epigenetic, and failing to recognize the importance of this factor in overall phenotype would be problematic as well. This is because epigenetic factors require an environmental influence to promote them, so that any editing to alter the potential for epigenetic expression would also have the capacity to produce unfortunate results.

There is an important message embedded in all the considerations we have just mulled over. Namely, that we would be profoundly unwise to hope that technology will fix our perceived shortcomings, or that evolution, for all of its past hyperactivity in our lineage, will ever ride in on its white horse and rescue us from our self-destructive tendencies and our myriad other deficiencies. Not only are the necessary preconditions entirely absent from the world we have created today, but there is no guarantee that, if something should improbably happen in this realm, it would affect the systems that govern our relationships with each other and with the environment around us. Accordingly, we are in all probability stuck with our imperfect, curve-bound biology; and if we don't like what we see, it is entirely up to us to change it. And the good news is that this is entirely possible. For, as we hope we have been able to show, one thing we do *not* have to worry about is the dismal prospect, proffered by the evolutionary psychologists, that our behaviors are a fixed burden that has been bequeathed to us by our ancestors, a burden that we shall have to wait for nature to evolve away at its own leisurely pace. To an amazing extent *Homo sapiens* makes its own environment, and the reaction norm ensures that with the appropriate social provisions we can adjust—if, that is, we can summon up both the will and the way, the former usually being much more difficult to find than the latter.

THE BELL CURVE AND THE HUMAN CONDITION

We began this book by noting that, for all their complexities and specificities, we cannot think of our behaviors as fixed items that are discrete and stand alone. As we have stressed, this is because, however we may choose to particularize them, all human behaviors actually lie on a spectrum. Our minds love to categorize and pigeonhole objects and ideas, but with few exceptions, in the real world we cannot approach any of our behaviors as if they were independent variables that we can define in isolation, or that we can hope to change without affecting anything else. The genome is simply too complex a place. And consequently, irrespective of whether we are therapists trying to sort out behavioral problems, or psychologists trying to understand the basis for human behaviors, or philosophers trying to use behavior as a means of gaining an insight into the more general human condition, it is usually rather unhelpful to attack the problem by singling out particular character traits for individual analysis. Invariably, it is the spectrum produced by the web of genetic causation that is most important to society and to the species as a whole. In social and evolutionary terms, what is critical is the curve itself.

Two characteristics of the normal distribution are especially important for society. The first of these is the mean value along the curve—how people behave on average. The greatest proportion of people in the population will mark the midpoint in the spectrum, and as we move away from the mean on either side we will find progressively fewer individuals toward each extreme. Of course, by itself the mean is purely descriptive, abstract, and typological. But if we were to find it less than ideal relative to some benchmark we have established, and we could somehow find some way of shifting it in the desired direction, we could pull the whole population along with it and, at least in principle, truly affect the nature of interaction within the society involved.

In the case of humans, both society and individual learning already exert this effect. Among social insects, the individual has little or no autonomy: a given chemical signal will reliably produce a specific behavioral result. But a huge proportion of human behaviors is learned, both from one's parents and from society at large. Indeed, an extended period of infant and juvenile learning and parental dependence is one of the most extraordinary characteristics of *Homo sapiens*. And precisely what an individual has learned will strongly affect how he or she will behave under a particular set of circumstances, even though that individual may theoretically have an almost unlimited menu of responses to choose from. As we saw earlier, human behavior thus broadly responds to the "norm of reaction," whereby the genotype specifies a range of possible outcomes, among which the environment—experience—selects. This is, of course, why any nature vs. nurture debate usually sums out as pretty fruitless when it comes to discussing human beings: both multiple genes and environment are normally involved, although their proportional contributions may vary. And as long as environment has any role at all, those averages have the potential to be shifted through the action of cultural stimuli—although there always remains the uncomfortable fact that, unless we can skew the distribution, half of it will inevitably linger to the negative side of the mean.

The second important characteristic of any curve is its shape. If something happens to elongate the curve to its sides, the outermost cases will be stretched away from each other so that the outliers will be even more extreme than they were before. And the farther apart the extremes, the greater is the potential for mutual incomprehension at best, and for outright conflict at worst. If, in contrast, the curve is compressed toward its center, individuals within the population will be more alike. Values around the mean will predominate, and the outliers will be reduced. And while under such circumstances the chances of conflict between the extremes might be somewhat diminished, the chances of tedium would almost certainly increase. Heuristically, it appears that the healthier and more viable curves are those that tend toward the classic bell form, with maximum symmetry and minimum expansion or compression from the normal distribution in which nearly all of the variation is symmetrically enclosed within two standard deviations of the mean.

Depending on the variable involved, what would be best for the social order often seems on the surface to be quite clear. Pretty much everyone would agree, for example, that society would be better off bending its efforts toward moving the mean to the right on the curve that describes the continuum from miserliness to generosity; and the same would also apply on the spectrum from violent to peaceable—unless, of course, that goal could be achieved only by the expedient of dulling our senses, a prospect we instinctively find appalling. But what about the continuum running from unimaginative to creative? In one sense it would be wonderful if we could maximize the creativity of the people in our communities by somehow shifting that curve; but if everyone were consequently highly strung and imaginative, the chances of conflict would surely shoot up. And although tedium would certainly be reduced, society would run the risk of rapidly becoming unlivable. So be careful what you wish for, because it's highly probable that the law of unintended consequences is the only true universal of human experience. As a result of that bell curve, the bad in our lives is the price we inevitably pay for the good; and promoting or eliminating one of them is almost certain to have unanticipated effects on the other.

Our voluntary behaviors are the product of a human psyche which, as we have seen, constructs its own particular mental world and then reacts to it. And it seems to do this in a holistic way, so that even describing human behaviors in terms of simple univariate spectra is an oversimplification. Whether we seek to define the human condition in a biological or a philosophical context, our definition has somehow to accommodate the sum total of human behaviors and experiences; and about the only way we have available to visualize or analyze this bewildering variety is to assemble all the different human behavioral spectra you could usefully imagine, each one a bell curve, and then to plot each in its own dimension.

Maybe you would first plot the frequency of all human individuals along an axis running from generosity to greed, then add a second axis running from unimaginative to artistic. All of that could theoretically be done on a flat sheet of paper. You could then add a vertical axis projecting straight up and down from your sheet of paper, and plot another curve running from, say, tongue-tied to eloquent. And while that is as much as any human being could visualize, statisticians have devised ways of adding variables in a conceptual hyperspace that could theoretically increase the number of its dimensions to include as many variables as you might want. You could then consider the human condition either to be all of that hyperspace or conceivably its centroid; but whichever you were to choose it would be an abstraction, impossible to relate to your or anyone else's human experience.

And it actually gets worse, complicating even theoretical ways of representing the human condition. For, the ever-nagging issue of value judgments aside, even such spectra as the one running from bad to good are in a very real sense abstractions when it comes to categorizing individuals. This is because individual human beings are amazingly labile in their behaviors. Although our general dispositions are usually set pretty early in our existences, our responses to changing stimuli mean that our reactions are inconsistent from one day to the next, and even from moment to moment. Not all of us are quite the extreme bundle of contradictions exemplified by Samuel Johnson's biographer James Boswell, to whom we alluded in our prologue; but we are all to some extent unpredictable because, regardless of external reality, the worlds we reconstruct in our minds are not only unique but are always changing, if usually only subtly.

If the individual human condition fluctuates in this way, how much more difficult is it to pinpoint our corporate human condition? Every descriptor we could imagine using for our species can be matched, somewhere within its ranks, by its own antithesis. To repeat Romain Gary's acute observation, "Part of being human is the inhumanity of it." The implacable reality embedded in Gary's insight places anyone trying to define—or even simply to characterize—the human condition in basically the same situation as United States Supreme Court justices wrestling with the issue of pornography. It is something they cannot define, even as they claim to know it when they see it.

SHIFTING THE CURVE

Whatever it really means to be human, a significant part of it seems to be a sort of existential discontent with ourselves, or with our lot, probably because we can always imagine that things *could* be better than they are—even if we have no idea how we might actually improve them. That dissatisfaction often translates into a yearning for change; and fortunately, the hope that we can somehow change ourselves is not an entirely forlorn one. Yes, barring a major disaster of some kind, any meaningful biological innovation in our lineage does indeed seem to be off the table for the foreseeable future. But in the cultural realm the prospects are very different. This is because, ever since our ancestors switched to the new symbolic cognitive algorithm, the history of humankind has revolved not around structural novelty, but around exploring something we already have: the new cultural capacity that was released at the moment of that change. And the desire for such exploration is apparently deeply embedded in the capacity itself.

As soon as symbolic human beings had arrived on the scene, the pattern of human activity documented in the archaeological record radically altered. The stasis typical of earlier times yielded—in what was, in evolutionary terms, an eyeblink—to the frenetic neophilia we see foreshadowed in the later Middle Stone Age of Africa. Symbolic thought, an entirely adventitious acquisition unshaped by selection, had brought with it the ability of each individual to remake the world in his or her mind. And although the idiosyncratic nature of this re-creation process has certainly led to a host of problems (if it's possible to think it, however crazy it might be, someone out there believes it or is even doing it), for the species as a whole this new predilection produced a previously unparalleled era in history, one characterized by a restless technological and intellectual probing of our cognitive limits, with old ideas and technologies going out of fashion and new ones being introduced at an increasing tempo.

That eventful history reflected not the serial acquisition of new capabilities as societies evolved, but rather the vigorous exploitation of the varied dimensions of a human biological potential that had already been established. New cultural features—art, religion, music, the wheel, the arch, writing systems, symphonies, nuclear bombs, computers, the Internet—were added piecemeal to the human behavioral repertoire by what amounted to a process of discovery. This process continues unabated, as cultures, technologies and individual beliefs diversify and develop. One interesting side effect of this exploration has been a consistent tendency to overcompensate. As new ideas have emerged, they have tended to be rejected on the one hand, and on the other adopted in all possible guises, up to and including being taken beyond their point of maximum utility and to the realm of crazy excess. Indeed, it is tempting to define humankind as the species that seems impelled to take every good idea to its ludicrous extremes. That bell curve again.

In the economic sphere, the exploration of our basic symbolic capacity has radically changed how we human beings relate to the world around us. And that relationship is still unstable: we are very far from having achieved any kind of new equilibrium with nature. Just how far away we are, or if such equilibrium is even possible, we have no idea. But we certainly have no reason whatever to suspect that we are currently anywhere close to exhausting this innovative capacity of ours. Indeed, we cannot even guess its limits. The only thing we can be entirely confident of is that there is still plenty to be explored in our unprecedented cognitive potential. This is why, despite superficial appearances, the prospect of biological stasis for our species in the millennia to come is very far from a recipe for future boredom. For the replacement of biological evolution by cultural evolution in human history is actually a very exciting thing, not least because things typically happen much faster in the cultural than in the biological realm. This swiftness stems ultimately from the fact that cultural innovations can be transmitted not simply from one generation to the next through time, but by horizontal transfer within generations. As a result, fads arise and disappear overnight, while more enduring novelties may with equal rapidity become established cultural properties. Culture may be defined in many ways, and in its narrowest definitions it is not uniquely a property of human beings. But in the way in which humans express it, culture *is* entirely unique; and in its human form it reveals vast vistas of possibility that no other species in the history of the planet has ever even come close to contemplating. In opening up those vistas, culture makes *Homo sapiens* not only an extraordinary natural experiment, but also an entirely unpredictable one. And anything is possible, because our "kluge-like" brains were never engineered by natural selection—or anything else—to produce a specific behavioral product.

So, how do things look? Well, the doomsayers have always been with us, and there has apparently never been a shortage of people around who consider contemporary life nothing but a precipitous cultural descent from a rosier past. An Assyrian clay tablet dating from 2800 BCE is said to have been inscribed with the following gloomy inscription: "Our Earth is degenerate in these latter days; there are signs that the world is speedily coming to an end; bribery and corruption are common; children no longer obey their parents; every man wants to write a book and the end of the world is evidently approaching." And whether or not this transcription is entirely accurate, or indeed even accurate at all, there can be little doubt that a significant portion of the human population (one tail of the bell curve) has always had a leaning toward making or believing such dispiriting prognostications, irrespective of their groundings in empirical fact or, perhaps more significantly, their lack of them. As recently as 2012, millions of Internet users from North America and around the world besieged conspiracy websites promoting the notion that the conclusion of a 5,126-year Mayan cycle on December 23, 2012, would mark the end of the world as we know it.

Even Sir Isaac Newton, the very epitome of the Age of Reason that witnessed the birth of modern science, was convinced that the end of the world was on its way. A huge amount of calculation based on biblical sources eventually revealed to him that this event would occur no sooner than 1,260 years after the foundation of the Holy Roman Empire. To the relief of most readers of this book, this will see us through at least 2060 CE; younger readers may also be comforted by Newton's caveat that "it may end later." Hedging of this sort was probably wise; around the end of the last century, the radio preacher Harold Camping was twice obliged by the passage of time to revise his predicted date for the coming Judgment Day.

Fortunately, no prediction of total catastrophe for humankind has yet been borne out by events, although human history has certainly had its ferocious ups and downs and pessimism has often proven to be in order. Humans have always had a better short-term grasp of cause and effect than long-term; and while wise people have usually seen local or global cataclysms coming, disaster has typically been allowed to follow anyway. This is, in fact, what appears to be happening right now in terms of climate change, in which humans are simultaneously exacerbating and ignoring a potentially disastrous phenomenon on the pretext that climate change is "natural" because it has happened in the past. But while rising sea levels, local aridification, amplified extreme weather, and so forth might not have mattered so much in the larger scheme of things when scattered bands of hunter-gatherers roamed the landscape, they clearly matter a very great deal in a period when enormous sedentary populations and a lot of critical infrastructure exist in places that are deeply threatened by their consequences. Scientists give us little more than a decade to reverse or alter today's warming trends or face irreversible adverse climatic effects; yet the political wind in the United States is blowing in precisely the opposite direction.

Still, although in this case the pessimists may have a higher chance of being proven right in the end, maybe more interesting are the optimists, the folks on the other end of the normal distribution who feel that things are actually getting better. The most widely read recent claim to this effect was made by the Harvard psychologist Steven Pinker in his 2011 book *The Better Angels of Our Nature: Why Violence Has Declined*. In that redoubtable tome Pinker argued that, despite the constant barrage of news about mayhem domestically and across the globe, the world is actually becoming a less violent place, and that it has indeed been on this track for a considerable period of time. According to Pinker, the rise of the modern nation-state, more extensive formal education, and the spread of abstract ideals such as those that underpin individual human rights, have all done their bit over the past few centuries to push humankind toward a more civilized way of doing business, both individually and among political entities.

In certain major respects, Pinker is undoubtedly correct. In those societies in which it has occurred, for example, the elimination of capital and corporal punishment has unquestionably helped promote more civil relations not only among individuals, but between the individual and the state. Similarly, the onslaught of humanitarian ideals has seen the suppression of slavery in the developed world. And in our society, the place an individual occupies on the bell curves of economic status and personal attainment is no longer entirely presaged by birth circumstance, so that women and members of minority groups can reasonably aspire to more desirable places on those curves than ever before. We live (despite our innate attraction to gloomy prophesies) in an age of unprecedented individual opportunity, material comfort, and freedom from disease; and it is unlikely that many of us would choose to trade our current circumstances for the more precarious ones that faced our ancestors.

Pinker attributes recent improvements in social and individual behaviors to the spread of Enlightenment ideas from the eighteenth century onward. And he argues on the political front that the development of larger and more powerful political entities, and of the increasingly complex apparatuses needed to run them, have also contributed to greater social order and general tranquility. It should be noted, though, that social improvements of this kind tend to come at some cost to individuals, who frequently chafe under the strictures involved; and indeed, the effectiveness of such measures is generally a function of how stringently they are enforced. Still, once again, the trick here is to find a balance in which society affords its members the protections they all—without exception—need without infringing too far on their personal freedoms. And determining "the greatest good of the greatest number" (or even what "good" itself means in this context) is hard to do without consulting those bell curves.

The bell curve itself may be inescapable, but what we might reasonably aspire to do is to shift its mean value. And Pinker does indeed give us good reason to believe that, at least in some areas of human experience, it is possible by cultural adjustment to move the mean for particular behaviors (child or spousal abuse, say, or homicide) in a desirable direction. Pinker suggests, for example, that murder rates are higher in the southern United States than in the northern areas of the continent because southerners are less willing than northerners are to tolerate a state monopoly on force. The resulting "self-help justice" leads inevitably to informal "culture of honor" sanctions among individuals. Get southerners to accept that it is the state's role to adjudicate quarrels, Pinker is saying, and homicide rates will drop. And it's certainly true that arbitrary measures aimed at changing people's behaviors or habits do sometimes work, and that they may even do so without excessive hardship to the individual. For example, on a more trivial level, but one that nonetheless had a major quality-of-life impact, since the passage of the "scoop the poop" law in New York City in 1978 a change in the attitudes of the citizenry has transformed the city's sidewalks away from the sanitary hazard they had been—and has consequently led to a generally happier populace. And sometimes, simple policy changes are just as effective as those more formally codified in the law: wherever in the developing world the education of girls has been prioritized, ages at first pregnancy have risen and skyrocketing rates of population growth have slowed.

A turn away from the current worldwide trend toward greater social and economic inequalities might help, too. A recent *Guardian* headline, "The Age of Entitlement: How Wealth Breeds Narcissism," showcased research by the American psychologist Paul Piff and his colleagues, who found by painstaking observation that drivers of expensive, high-status cars were three times less likely than drivers of more lowly vehicles to yield to pedestrians at crosswalks, and four times more likely to inconsiderately cut off their more courteous fellow citizens who drive inexpensive old cars. This finding will not surprise anyone living in Germany, where drivers of top-of-the-line BMWs are almost ten times as likely to pick up driving convictions as are those who drive a modest Ford Ka. And neither will it surprise any regular user of the New York City subway, who will almost certainly have noticed that more affluently dressed people are much less likely than those more cheaply attired to give to panhandlers. This informal observation is fully in line with studies showing that Americans in the top 20 percent earnings bracket typically donate 1.3 percent of their income to charity, while those in the bottom 20 percent donate 3.2 percent of what they earn (without any hope of getting their name on a building). Piff followed up his initial street-side observations with structured laboratory investigations that revealed his wealthier students to be more likely to contemplate "stealing or benefiting from things to which they were not entitled" than their less economically advantaged fellows were. They were also more likely to agree with such statements as "I am more deserving than other people," an attitude that is clearly reflected in recent adjustments to the tax laws in the United States. All in all, Piff and colleagues wrote, "relative to lower-class individuals, individuals from upper-class backgrounds behaved more unethically in both naturalistic and laboratory settings."

In a world that measures its success by the size of its gross domestic product, this might look like inescapably bad news: as national wealth increases so, it seems, will unethical behaviors. But as he dug deeper, Piff actually found that behavior is highly responsive to circumstance, and that even the rich could behave more empathetically when shown, say, pictures of children in economic distress. The lesson appears to be that it is not necessarily bad people who become rich, or that rich people are inherently bad; but that wealth, and the isolation it provides from the less edifying realities of life, actively induces those feelings of entitlement and uncaring. In turn, this suggests that reducing income disparities through, say, strongly progressive taxation, might not simply promote a sectionally desirable goal (remember the monkey's cucumber and France's *gilets jaunes*), but that it might actually increase the prevalence of the kind of socially responsible behaviors that most members of society would consider worthwhile. Currently, things appear to be heading energetically in the contrary direction; but, as inequalities mount all around us, the journalist and social philosopher Anne Manne urges us in her book *The Life of I* not to forget the example of Ayn Rand, the neoconservative darling and author of such bestsellers as *The Virtue of Selfishness.* After decades spent lauding society's wealthy "producers" and railing against less-advantaged "parasites," Rand eventually signed up for Social Security and Medicare.

The idea that, with the appropriate social incentives, attitudes and behaviors can indeed be made to change for the better finds some really surprising support in the austerely beautiful and tragic country of Yemen, where gun ownership per capita has traditionally been second only to that of the United States of America, and where many a male would feel distinctly underdressed if he went out in public without a Kalashnikov slung over his shoulder in addition to the fearsome *jambiya* dagger stuck in his belt. Yet in the southern port city of Mukalla, the only arms visible on the street today are carried by soldiers, a reminder not only of the externally-sponsored civil war that has torn Yemen cruelly apart over the past several years, but more specifically of the town's brief takeover in 2015 by al-Qaeda. A huge influx of weapons into the town followed this incursion, as forces backed by the United Arab Emirates battled to oust the jihadists. But once a semblance of normality had been restored, the army imposed a ban on carrying weapons in public—and today visitors to the city literally have to check their guns at the door. This draconian move could never have succeeded anywhere else in Yemen; but with strong trading ties across the Indian Ocean, the legacy of a British occupation during which tribal identities and feuds were suppressed, and the local popularity of the peaceable Sufi strain of Islam, Mukalla has evidently become culturally different from its hinterland. A resident of the city was even recently quoted by *The Economist* as saying, "We don't like carrying guns." Anywhere else in Yemen this sentiment might be taken as denying manhood itself. Larger influences really can change human attitudes.

Still, one thing that nonetheless doesn't seem to have changed in all the wide range of human experience that Pinker surveys is the underlying distribution of human behaviors. Within many (though not all) societies, interindividual violence has been contained. But it has not been eliminated anywhere, and it could be argued that on the level of societies as a whole, things have actually become rather worse. For it is the state that goes to war, and although with fewer and larger entities there will presumably be diminishing numbers of conflicts, when large nation-states do come to blows, the scale of the ensuing violence not only inevitably increases, but tends to ratchet up with time, as a function of technological advancement that is often spurred by the violence itself. The lesson here seems to be that, even in those societies that have become more orderly, the basics of human nature haven't changed since historical records began to be kept—and presumably since long before that. The pattern persists, even if the mean changes.

So, in the end the yin and the yang—those opposite tails of the normal distribution—seem to be omnipresent and unavoidable, no matter what area of human experience one looks at. They will never go away; one will always imply the existence of the other; and their expression will always somehow be spread more or less equally around the mean. This strongly implies, to repeat an earlier observation, that what we perceive as the bad in our lives is the price that we inevitably pay for the good, just as the irrational is nature's inevitable tribute to the lucid. On the social front, there will never be rich people without the poor—something the rich will forget to their ultimate disadvantage as the curve gets more and more Paretian. And the sum total of happiness in the world will continue to be approximately the same as the sum total of misery, though one might hope for periods in which that whole curve might be shifted to the left. In the end, the human comedy is equally the human tragedy; and far too often the hapless individual human being will have little control over the kind of drama he or she will be acting in.

Still, Pinker has piled up an impressive mass of documentation in defense of his thesis, the corollary of which is that, in human societies, both the shape of the curve for any given behavior and the position of its mean are indeed responsive to stimuli such as evolving abstract ideas, changing cultural norms, new legislation, and more effective practical sanctions or encouragements. This flexibility inevitably cuts both ways, as the current coarsening of our politics and society clearly demonstrates; but it does remain true that, if we truly want to, we *can* in principle move those bell curves—although whether they can always be moved without materially infringing on individual freedoms is another matter. And therein lies the ultimate inconsistency of human experience: human beings are intensely social and reproductively successful animals who just happen to tolerate living in very high densities, while at the same time the emergent quality of each individual's consciousness assures that the world he or she lives in is not quite identical to anybody else's. The resulting clash between individual desires and the elaborate social institutions necessary to keep complex human societies functioning is one of the major prices we pay for having our unprecedented cognitive powers. Free will—the accidental product of the emergent origin of modern human consciousness—is a unique gift of Mother Nature to *Homo sapiens*. But it comes at a cost that is more than purely existential.

Those particular tensions exist, of course, at the level of society; and the need to resolve them will keep philosophers, sociologists, politicians, and many others busy as long as there are people on the planet. But, like corporations and even armies, societies do not feel pain. The only unit of true suffering in this world is the individual; and any society that forgets this basic truth will be condemned to increase the sum total of human misery within it. Politicians take note. But it is equally true that it is only individuals who have ideas, and desires, and ambitions. Only the individual experiences joy, exhilaration, and the satisfaction of achievement. Only the individual feels love and attachment. Only the individual can create a vision for his or her own future, or for the future of some larger entity. Only the individual truly feels the fear of impoverishment, or of frailty, or of death. Only the individual can make the conscious decision to moderate his or her more antisocial urges. And accordingly, it is only individuals, whether working collectively or on their own, who will ultimately make the difference to the quality of their own lives and those of others. And, as tiny as the individual may seem in relation to vast faceless societies or organizations, this knowledge is actually empowering. With respect to behavioral distributions, an individual is only one small data point in a larger collection of data that makes up any bell curve. But, as infinitesimal as the effect of each individual might appear to be, it is in the end the individual who changes the shape of the curve, and thereby ultimately the entire trajectory of human behaviors. An individual chipped the first stone. An individual engraved the first symbolic object. An individual invented the flushing toilet. And if everything has to start with an individual, then everyone can play a role in the ongoing enterprise that is *Homo sapiens*, secure in the knowledge that they are free from the tyranny of an imagined biological past.

NOTES AND BIBLIOGRAPHY

ONE: GENES, EVOLUTION, AND THE BELL CURVE

Sir Francis Crick's famous quote can be found in his book *The Astonishing Hypothesis* (Crick, 1994). Søren Eilers's clever Lego counting was published in Eilers (2016). For a more detailed description of the biology and history of DNA and genomics the reader can consult many textbooks, but we list one by the authors here (DeSalle and Yudell, 2019). Mayr's (1964) original reference to beanbag genetics and Haldane's (1964) rebuttal are interesting places to start examination of the role of genetics in evolutionary theory. Deborah Skinner's interaction with the Skinner Box is described in Benjamin, et al. (1999) and identical twins Scott and Mark Kelly's differences after space travel are described in Hoshide and Jandial (2018). Danielle Posthuma and colleagues' analysis of five decades of twin research is summarized in Polderman, et al. (2015). We have discussed several aspects of the history behind the study of evolution, and these can be found in DeSalle and Tattersall (2018) and Tattersall and DeSalle (2011). Chevalier de Méré's gambling problem is discussed in many introductory statistics books including Feller (1968), and Abraham de Moivre's penchant for distributions is described by Bellhouse (2011). The development of the normal curve and its somewhat twisted history are described by Stahl (2006), and Blakeslee's normal distribution of University of Connecticut men can be found in several places on the web, one of which we list below. A discussion of the shortcomings of the normal distribution in studying humans is given in Robinson (2017), and a clear description of the Paretian distribution is given by O'Boyle and Aguinas (2012). The development of the new synthesis is described by Mayr (1964). See Gould and Lewontin (1979) for spandrels.

Bellhouse, D. R. 2011. *Abraham De Moivre: Setting the Stage for Classical Probability and Its Applications*. London: Taylor & Francis.

Benjamin Jr., L. T., and E. Nielsen-Gammon. 1999. "B. F. Skinner and Psychotechnology: The Case of the Heir Conditioner." *Review of General Psychology* 3 (3): 155–160.

Blakeslee Distribution. http://advance.uconn.edu/1999/990201/020199hs.htm.

Crick, F. 1994. *The Astonishing Hypothesis: The Scientific Search for the Soul*. New York: Scribner.

DeSalle, R., and I. Tattersall. 2019. *Troublesome Science: The Misuse of Genetics and Genomics in Understanding Race*. New York: Columbia University Press.

DeSalle, R., and M. Yudell. 2019. *Welcome to the Genome: A User's Guide to the Genetic Past, Present, and Future*, 2nd ed. Hoboken, NJ: Wiley.

Eilers, S. 2016. "The LEGO Counting Problem." *The American Mathematical Monthly* 123 (5): 415–426.

Feller, W. 1968. *An Introduction to Probability Theory and Its Applications*, vol. 1. New York: Wiley.

Gould, S. J., and R. C. Lewontin. 1979. "The Spandrels of San Marco and the Panglossian Paradigm: A Critique of the Adaptationist Programme." *Proceedings of the Royal Society of London, B* 205 (1161): 581–598.

Haldane, J. B. S. 1964. "A Defense of Beanbag Genetics." *Perspectives in Biology and Medicine* 7 (3): 343–360.

Hoshide, R., and R. Jandial. 2018. "The Genetics of Space Travel." *Neurosurgery* 83 (1): E8–E9.

Mayr, Ernst. 1964. *Animal Species and Evolution*. Cambridge, MA: Belkap Press.

O'Boyle Jr., E., and H. Aguinis. 2012. "The Best and the Rest: Revisiting the Norm of Normality of Individual Performance." *Personnel Psychology* 65 (1): 79–119.

Polderman, T. J. C., B. Benyamin, C. A. De Leeuw, et al. 2015. "Meta-analysis of the Heritability of Human Traits Based on Fifty Years of Twin Studies." *Nature Genetics* 47 (7): 702–710.

Robinson, L. R. 2017. "It's Time to Move On from the Bell Curve." *Muscle & Nerve* 56 (5): 859–860.

Stahl, S. 2006. "The Evolution of the Normal Distribution." *Mathematics Magazine* 79 (2): 96–113.

Tattersall, I., and R. DeSalle. 2011. *Race? Debunking a Scientific Myth*. College Station, TX: Texas A&M University Press.

TWO: SCIENCE AND BEHAVIOR: TRAPPED BETWEEN SIMPLICITY AND COMPLEXITY

Edward O. Wilson's tome *Sociobiology* was published first in 1975 and was reissued on its twenty-fifth anniversary in 2000. Richard Lewontin's 1974 masterpiece is listed below, as well as his review articles that were seminal in developing the field of molecular population genetics (Lewontin 1967, 1970, 1973, 1985, 2002). Richard Dawkins's groundbreaking book *The Selfish Gene*, originally published in 1976, is also listed below, and although

Dawkins has not articulated his ideas about homosexuality in print, we reference a video of him explaining them over coffee. Trudy Mackay's work on *Drosophila*, the allelic spectrum, and the GWAS quadrant system are summarized by Mackay (2009), Reich and Lander (2001), and McCarthy, et al. (2008), respectively. GWAS Central can be found on the web at https://www.gwascentral.org/. PKU as an example of missing heritability was originally discussed by Vineis and Pearce (2010), and omnigenics is described in Boyle, et al. (2017). Matthew Rockman's papers discussed in this chapter are Rockman (2012) and Paaby and Rockman (2013). The molecular basis of melanism in the rock pocket mouse was first explained by Nachman, et al. (2003), and it was expanded upon later by Hoekstra and colleagues (Hoekstra and Nachman, 2003; Hoekstra, et al., 2004; Hoekstra, et al., 2005). The hopeful monster (Hawaiian *Drosophila* with modified mouthparts) is described in DeSalle and Carew (1992). Facial recognition and landmarks are discussed by Claes, et al. (2014, 2018; Shaffer, et al., 2016; Lippert, et al., 2017; Cole, et al., 2016), and GWAS of height is discussed in Wood, et al. (2014). The Bateson quote can be found in Rockman (2012). Jens Christian Clausen, David Keck, and William Heisey's work is described in *The Triple Helix* (Lewontin, 2001), and in a wonderful review article by Núñez-Farfán and Schlichting (2001), as well as in the original article by the trio. More on reaction norms can be found in Gomulkiewicz and Kirkpatrick (1992). Weiss quotes are from Buchanan, et al. (2009) and Weiss and Buchanan (2011).

Boyle, E. A., Y. I. Li, and J. K. Pritchard. 2017. "An Expanded View of Complex Traits: From Polygenic to Omnigenic." *Cell* 169 (7): 1177–1186.

Buchanan, A. V., S. Sholtis, J. Richtsmeier, and K. M. Weiss. 2009. "What Are Genes 'For' or Where Are Traits 'From'? What Is the Question?" *Bioessays* 31 (2): 198–208.

Claes, P., D. K. Liberton, K. Daniels, K. M. Rosana, E. E. Quillen, L. N. Pearson, B. McEvoy, et al. 2014. "Modeling 3D Facial Shape from DNA." *PLoS Genetics* 10 (3): e1004224.

Claes, P., J. Roosenboom, J. D. White, T. Swigut, D. Sero, J. Li, M. K. Lee, et al. 2018. "Genome-wide Mapping of Global-to-Local Genetic Effects on Human Facial Shape." *Nature Genetics* 50 (3): 414–422.

Clausen, J., and Heisey, W. M. 1958. *Experimental Studies on the Nature of Species, Volume IV: Genetic Structure of Ecological Races.* Publication 615. Washington, DC: Carnegie Institution of Washington.

Cole, J. B., M. Manyama, J. R. Larson, D. K. Liberton, T. M. Ferrara, S. L. Riccardi, M. Li, et al. 2016. "Human Facial Shape and Size Heritability and Genetic Correlations." *PLoS Genetics* 12 (8): e1006174.

Dawkins, R. 2006. *The Selfish Gene: With a New Introduction by the Author.* Oxford, UK: Oxford University Press. (Originally published in 1976).

Dawkins, Richard. An explanation for the persistence of homosexuality. https://www.huffingtonpost.com/2011/11/06/dawkins-evolution-homosexuality_n_1078714.html.

DeSalle, R., and E. Carew. 1992. "Phyletic Phenocopy and the Role of Developmental Genes in Morphological Evolution in the Drosophilidae." *Journal of Evolutionary Biology* 5 (3): 363–374.

Gomulkiewicz, R., and M. Kirkpatrick. 1992. "Quantitative Genetics and the Evolution of Reaction Norms." *Evolution* 46 (2): 390–411.

Hoekstra, H. E., and M. W. Nachman. 2003. "Different Genes Underlie Adaptive Melanism in Different Populations of Rock Pocket Mice." *Molecular Ecology* 12 (5): 1185–1194.

Hoekstra, H. E., K. E. Drumm, and M. W. Nachman. 2004. "Ecological Genetics of Adaptive Color Polymorphism in Pocket Mice: Geographic Variation in Selected and Neutral Genes." *Evolution* 58 (6): 1329–1341.

Hoekstra, H. E., J. G. Krenz, and M. W. Nachman. 2005. "Local Adaptation in the Rock Pocket Mouse (*Chaetodipus intermedius*): Natural Selection and Phylogenetic History of Populations." *Heredity* 94 (2): 217–225.

Lewontin, R. C. 1967. "Population Genetics." *Annual Review of Genetics* 1: 37–70.

Lewontin, R. C. 1970. "The Units of Selection." *Annual Review of Ecology and Systematics* 1: 1–18.

Lewontin, R. C. 1973. "Population Genetics." *Annual Review of Genetics* 7: 1–17.

Lewontin, R. C. 1974. *The Genetic Basis of Evolutionary Change*. New York: Columbia University Press.

Lewontin, R. C. 1985. "Population Genetics." *Annual Review of Genetics* 19: 81–102.

Lewontin, R. C. 2001. *The Triple Helix: Gene, Organism, and Environment*. Cambridge, MA: Harvard University Press.

Lewontin, R. C. 2002. "Directions in Evolutionary Biology." *Annual Review of Genetics* 36: 1–18.

Lippert, C., R. Sabatini, M. C. Maher, E. Y. Kang, S. Lee, O. Arikan, A. Harley, et al. 2017. "Identification of Individuals by Trait Prediction Using Whole-genome Sequencing Data." *Proceedings of the National Academy of Sciences, USA* 114 (38): 10166–10171.

Mackay, T. F. C. 2009. "The Genetic Architecture of Complex Behaviors: Lessons from *Drosophila*." *Genetica* 136 (2): 295–302.

McCarthy, M. I., G. R. Abecasis, L. R. Cardon, D. B. Goldstein, J. Little, J. P. A. Ioannidis, and J. N. Hirschhorn. 2008. "Genome-wide Association Studies for Complex Traits: Consensus, Uncertainty, and Challenges." *Nature Reviews Genetics* 9 (5): 356–362.

Nachman, M. W., H. E. Hoekstra, and S. L. D'Agostino. 2003. "The Genetic Basis of Adaptive Melanism in Pocket Mice." *Proceedings of the National Academy of Sciences, USA* 100 (9): 5268–5273.

Núñez-Farfán, J., and C. D. Schlichting. 2001. "Evolution in Changing Environments: The 'Synthetic' Work of Clausen, Keck, and Heisey." *The Quarterly Review of Biology* 76 (4): 433–457.

Paaby, A. B., and M. V. Rockman. 2013. "The Many Faces of Pleiotropy." *Trends in Genetics* 29 (2): 66–73.

Reich, D. E., and E. S. Lander. 2001. "On the Allelic Spectrum of Human Disease." *TRENDS in Genetics* 17 (9): 502–510.
Rockman, M. V. 2012. "The QTN Program and the Alleles That Matter for Evolution: All That's Gold Does Not Glitter." *Evolution: International Journal of Organic Evolution* 66 (1): 1–17.
Shaffer, J. R., E. Orlova, M. K. Lee, E. J. Leslie, Z. D. Raffensperger, C. L. Heike, M. L. Cunningham, et al. 2016. "Genome-wide Association Study Reveals Multiple Loci Influencing Normal Human Facial Morphology." *PLoS Genetics* 12 (8): e1006149.
Vineis, P., and N. Pearce. 2010. "Missing Heritability in Genome-wide Association Study Research." *Nature Reviews Genetics* 11 (8): 589–600.
Weiss, K. M., and A. V. Buchanan. 2011. "Is Life Law-like?" *Genetics* 188 (4): 761-771.
Wilson, E. O. 2000. *Sociobiology*, Anniversary ed. Cambridge, MA: Harvard University Press.
Wood, A. R., T. Esko, J. Yang, S. Vedantam, T. H. Pers, S. Gustafsson, A. Y. Chu, et al. 2014. "Defining the Role of Common Variation in the Genomic and Biological Architecture of Adult Human Height." *Nature Genetics* 46 (11): 1173–1180.

THREE: EMERGENCE OF THE HUMAN COGNITIVE STYLE

A general consideration of human cognitive evolution can be found in Tattersall (2012), and a survey of how we came to know what we think we know of the subject is in Tattersall (2015a). Both books are extensively referenced. Begun (2015) presents a fine survey of the fossil ape radiation. Gibbons (2007) provides an entertaining account of the discovery of the very early hominids, and the special edition of *Science* on *Ardipithecus* (White, et al. 2009, et seq.) provides technical detail. An overview of australopith species is given by Kimbel (2015), and isotopic evidence for australopith diets is reviewed by Sponheimer, et al. (2013). Kappelman, et al. (2016) suggest that Lucy died falling from a tree. Hart and Sussman (2009) present an iconoclastic and persuasive view of early hominid social organization. Klein (2009) remains the best general review of both early and later phases of stone tool making and other Paleolithic technologies. Harcourt-Smith (2015) reviews early hominid locomotor adaptations, and Wrangham (2009) emphasizes the significance of fire domestication. Tattersall (2015b and 2016, respectively) overviews *Homo ergaster* and the genus *Homo*. For a look at *Homo heidelbergensis* and *Homo neanderthalensis*, consult Stringer (2013). For Neanderthal environments and adaptations, see van Andel and Davies (2003). Massacred Neanderthals at El Sidrón and their DNA were reported by Lalueza-Fox, et al. (2010), and cannibalism among middle Pleistocene European hominids is discussed by Carbonell, et al. (2010). Staubwasser, et al. (2018) suggest that Neanderthals were under extreme climatic stress around the time of their demise. Hoffmann, et al. (2018) reported putative Neanderthal cave decoration, while Slimak, et al. (2018) argued for a more recent date. See Bar-Yosef and Bordes (2010) and Higham, et al. (2010) for discussion of the Châtelperronian. Very early *Homo sapiens* in Ethiopia were reported by White, et al. (2003) and McDougall, et al. (2005). Early symbolic manifestations in Africa have been summarized by Dubreuil and Henshilwood and other contributors to Lefebvre, Comrie, and Cohen (2013). Fire-hardening technology in the MSA was first reported by Brown, et al. (2009), and Marean (2014) discussed early economic indicators of symbolic cognition. DeSalle (2018) overviewed the DNA evidence for early human movements out of Africa and across the globe. See von Petzinger (2016) for a review of early symbolic behaviors in Europe, and Aubert, et al. (2014) for the report of early animal imagery in Sulawesi. Hinzen quote is from Hinzen (2012). For algorithmic basis of language see Berwick and Chomsky (2017), and for spontaneous language invention see Senghas, Kita, and Özyürek (2005). Tattersall (2018) reviews the tricky relationship between brain size and cognitive mode.

Aubert M., A. Brumm, M. Ramli, T. Sutikna, et al. 2014. "Pleistocene Cave Art from Sulawesi, Indonesia." *Nature* 514: 223–227.

Bar-Yosef, O., and Bordes, J.-G. 2010. "Who Were the Makers of the Châtelperronian Culture?" *Journal of Human Evolution* 59: 686–593.

Begun, D. R. 2015. *The Real Planet of the Apes: A New Story of Human Origins*. Princeton, NJ: Princeton University Press.

Berwick, R. C., and N. Chomsky. 2017. *Why Only Us: Language and Evolution*. Cambridge, MA: MIT Press.

Brown, K. S., et al. 2009. "Fire As an Engineering Tool of Early Modern Humans." *Science* 325: 859–862.

Carbonell, E., I. Caceres, M. Lizano, P. Saladie, et al. 2010. "Cultural Cannibalism as a Paleoeconomic System in the European Lower Pleistocene." *Current Anthropology* 51: 539–549.

DeSalle, R. 2018. "The Paleogenomic Revolution: New Bearings in Human Dispersal." *Natural History* 126 (8): 23–26.

Gibbons, A. 2007. *The First Human: The Race to Discover Our Earliest Ancestors*. New York: Anchor Books.

Harcourt-Smith, W. E. H. 2015. "The Origins of Bipedal Locomotion." In *Handbook of Paleoanthropology*, 2nd ed., Vol. 3, W. Henke and I. Tattersall, eds. Heidelberg: Springer, pp. 1919–1960.

Hart, D., and R. W. Sussman. 2009. *Man the Hunted: Primates, Predators, and Human Evolution,* expanded edition. Boulder, CO: Westview Press.

Higham T., R. Jacobi, M. Julien, F. David, et al. 2010. "Chronology of the Grotte du Renne (France) and Implications for the Context of Ornaments and Human Remains within the Châtelperronian." *Proceedings of the National Academy of Sciences, USA* 107: 20234–20239.

Hinzen, W. 2012. "The Philosophical Significance of Universal Grammar." *Language Sciences* 34: 635–649.

Hoffmann, D. L., C. D. Standish, M. Garcia-Diez, P. B. Pettit, et al. 2018. "U-Th dating of Carbonate Crusts Reveals Neandertal Origin of Iberian Cave Art." *Science* 359: 912–915.

Kappelman, J., R. A. Ketcham, S. Pearce, L. Todd, et al. 2016. "Perimortem Fractures in Lucy Suggest Mortality from Fall Out of Tall Tree." *Nature* 537: 503–508.

Kimbel, W. 2014. "The Species and Diversity of Australopiths." In *Handbook of Paleoanthropology*, 2nd ed., Vol. 3, W. Henke and I. Tattersall, eds. Heidelberg: Springer, pp 1539–1573.

Klein, R. 2009. *The Human Career: Human Biological and Cultural Origins*, 3rd ed. Chicago: University of Chicago Press.

Lalueza-Fox, C., A. Rosas, A. Estalrrich, E. Gigli, et al. 2010. "Genetic Evidence for Patrilocal Mating Behavior Among Neandertal Groups." *Proceedings of the National Academy of Sciences, USA* 108 (1): 250–253.

Lefebvre, C., B. Comrie, and H. Cohen. 2013. *New Perspectives on the Origin of Language*. Amsterdam/Philadelphia: John Benjamins.

Marean C. 2014. "The Origins and Significance of Coastal Resource Use in Africa and Western Asia." *Journal of Human Evol*ution 77: 17–40.

McDougall, I., F. H. Brown and J. G. Fleagle. 2005. "Stratigraphic Placement and Age of Modern Humans from Kibish, Ethiopia." *Nature* 433: 733–736.

Senghas A., S. Kita, A. Özyürek. 2005. "Children Creating Core Properties of Language: Evidence from an Emerging Sign Language in Nicaragua." *Science* 305: 1779–1782.

Slimak, L., J. Fietzke, J.-M. Geneste, R. Ontanon, R. 2018. Comment on "U-Th Dating of Carbonate Crusts Reveals Neandertal Origin of Iberian Cave Art." *Science* 361: 912–915.

Sponheimer, M., Z. Alemseged, T. E. Cerling, F. E. Grine, et al. 2013. "Isotopic Evidence of Early Hominin Diets." *Proceedings of the National Academy of Sciences, USA* 110 (26): 10513–10518.

Staubwasser. M., V. Dragusin, B. Onac, S. Assonov, et al. 2018. "Impact of Climate Change on the Transition of Neanderthals to Modern Humans in Europe." *Proceedings of the National*

Academy of Sciences, USA, www.pnas.org/cgi/doi/10.1073/pnas.1808647115.

Stringer, C. B. 2013. *Lone Survivors: How We Came to Be the Only Humans on Earth.* New York: St. Martin's Griffin.

Tattersall, I. 2012. *Masters of the Planet: The Search for Our Human Origins.* New York: Palgrave Macmillan.

Tattersall, I. 2015a. *The Strange Case of the Rickety Cossack, and Other Cautionary Tales from Human Evolution.* New York: Palgrave Macmillan.

Tattersall, I. 2015b. "*Homo ergaster* and Its Contemporaries." In *Handbook of Paleoanthropology,* 2nd ed., Vol. 3, W. Henke and I. Tattersall, eds. Heidelberg: Springer, pp. 2167–2188.

Tattersall, I. 2016. "The Genus *Homo.*" *Inference: International Review of Science* 2 (1): http://inference-review.com/article/the-genus-homo.

Tattersall, I. 2018. "Brain Size and the Emergence of Modern Human Cognition." In J. H. Schwartz, ed., *Rethinking Human Evolution.* Cambridge, MA: MIT Press, pp. 319–334.

Van Andel, T. H., and W. Davies, eds. 2003. *Neanderthals and Modern Humans in the European Landscape During the Last Glaciation: Archaeological Results of the Stage 3 Project.* Cambridge, M.A.: McDonald Institute for Archaeological Research.

Von Petzinger, G. 2016. *The First Signs: Unlocking the Mysteries of the World's Oldest Symbols.* New York: Atria Books.

White, T. D., B. Asfaw, D. DeGusta, H. Gilbert, G. D. Richards, G. Suwa, and F. C. Howell. 2003. "Pleistocene *Homo sapiens* from Middle Awash, Ethiopia." *Nature* 423: 742–747.

White, T. D., B. Asfaw, Y. Beyene, Y. Haile-Selassie, C. O. Lovejoy, G. Suwa, and G. Woldegabriel. 2009. "*Ardipithecus ramidus* and the Paleobiology of Early Hominids." *Science* 326 (5949): 75–86.

Wrangham, R. 2009. *Catching Fire: How Cooking Made Us Human.* New York, Basic Books.

FOUR: GENES, PEOPLE, AND BEHAVIOR

Homosexual giraffes are discussed in Coel (1967), Ito et al. (1996), and Grosjean, et al. (2008). Wilson's *Sociobiology* was referenced in Chapter 2. Michael Ghiselin first mentioned evolutionary psychology in 1973, and Leda Cosmides and John Tooby's primer of evolutionary psychology can be found online as listed below. Gary Marcus's wonderful treatment of the messiness of the human brain can be found in his 2009 book *Kluge* and David Eagleman's description of the "team of rivals" is in his 2013 book *Incognito*, both referenced below. A description of split brain and Capgras syndromes and their possible neural interpretations can be found in Gazzaniga (2005) and Hirstein and Ramachandran (1997), respectively. The God Helmet is discussed in Persinger, et al. (2010), and Hamer's God gene and Zimmer's characterization of it are also referenced below. References for each of the behaviors we address in the text are given below under their section title.

Coel, Malcolm J. 1967. "'Necking' Behaviour in the Giraffe." *Journal of Zoology* 151 (1): 313–321.

Cosmides, L. and J. Tooby. *Primer on Evolutionary Psychology*, (https://www.cep.ucsb.edu/primer.html).

Eagleman, David. 2013. *Incognito*. Paris: Robert Laffont.

Ghiselin, M. T. 1973. "Darwin and Evolutionary Psychology." *Science* 179 (4077): 964–968.

Grosjean, Y., M. Grillet, H. Augustin, J.F. Ferveur, and D. E. Featherstone. 2008. "A Glial Amino-acid Transporter Controls Synapse Strength and Courtship in *Drosophila*." *Nature Neuroscience* 11 (1): 54.

Hirstein, W. and V. S. Ramachandran. 1997. "Capgras Syndrome: A Novel Probe for Understanding the Neural Representation of the Identity and Familiarity of Persons." *Proceedings of the Royal Society of London B: Biological Sciences* 264 (1380): 437–444.

Ito, H., K. Fujitani, K. Usui, K. Shimizu-Nishikawa, S. Tanaka, and D. Yamamoto. 1996. "Sexual Orientation in *Drosophila* Is Altered by the Satori Mutation in the Sex-determination Gene Fruitless That Encodes a Zinc Finger Protein with a BTB Domain." *Proceedings of the National Academy of Sciences, USA* 93 (18): 9687–9692.

Marcus, G. 2009. *Kluge: The Haphazard Evolution of the Human Mind.* New York: Houghton Mifflin Harcourt.

Gazzaniga, M. S. 2005. "Forty-five Years of Split-brain Research and Still Going Strong." *Nature Reviews Neuroscience* 6 (8): 653.

RELIGION

Hamer, D. H. 2005. *The God Gene: How Faith Is Hardwired into our Genes.* New York: Anchor.

Persinger, M. A., K. S. Saroka, S. A. Koren, and L. S. St-Pierre. 2010. "The Electromagnetic Induction of Mystical and Altered States Within the Laboratory." *Journal of Consciousness Exploration & Research* 1 (7): 808–830.

Zimmer, C. 2004. "Faith-boosting Genes." *Scientific American* 291 (4): 110–111.

ETHICS, RACISM, AND WARFARE

Gottschalk, Martin, and Lee Ellis. 2009. "Evolutionary and Genetic Explanations of Violent Crime." In *Violent Crime: Clinical and Social Implications.* Thousand Oaks, CA: Sage, pp. 57–74.

Potts, M. 2010. *Sex and War: How Biology Explains Warfare and Terrorism and Offers a Path to a Safer World.* Dallas, TX: BenBella Books. Kindle Edition, pp. 386–387.

Veroude, K., Y. Zhang-James, N. Fernàndez-Castillo, M. J. Bakker, B. Cormand, and S. V. Faraone. 2016. "Genetics of Aggressive

Behavior: An Overview." *American Journal of Medical Genetics Part B: Neuropsychiatric Genetics* 171 (1): 3–43.

Wagner, R. T. 2017. "The Significant Influencing Factors of Xenophobia." *Student Scholarship—Education* 2. https://digitalcommons.olivet.edu/educ_stsc/2.

POLITICS

Fehr, E., and S. Gächter. 2002. "Altruistic Punishment in Humans." *Nature* 415 (6868): 137–143.

Hatemi, P. K., N. A. Gillespie, L. J. Eaves, B. S. Maher, B. T. Webb, A. C. Heath, S. E. Medland, et al. 2011. "A Genome-wide Analysis of Liberal and Conservative Political Attitudes." *The Journal of Politics* 73 (1): 271–285.

King, A. J., D. D. P. Johnson, and M. Van Vugt. 2009. "The Origins and Evolution of Leadership." *Current Biology* 19 (19): R911–R916.

Lockyer, A., and P. K. Hatemi. 2018. "Genetics and Politics: A Review for the Social Scientist." In *The Oxford Handbook of Evolution, Biology, and Society*. Oxford: Oxford University Press, pp. 281–291.

Van Vugt, Mark, and R. Ronay. 2014. "The Evolutionary Psychology of Leadership: Theory, Review, and Roadmap." *Organizational Psychology Review* 4 (1): 74–95.

SEXUAL BEHAVIORS

Bailey, J. M., and R. C. Pillard. 1991. "A Genetic Study of Male Sexual Orientation." *Archives of General Psychiatry* 48 (12): 1089–1096.

Bailey, J. M., R. C. Pillard, M. C. Neale, and Y. Agyei. 1993. "Heritable Factors Influence Sexual Orientation in Women." *Archives of General Psychiatry* 50 (3): 217–223.

Hellmer, K., T. J. Stensson, and K. M. Jylhä. 2018. "What's (Not) Underpinning Ambivalent Sexism? Revisiting the Roles of

Ideology, Religiosity, Personality, Demographics, and Men's Facial Hair in Explaining Hostile and Benevolent Sexism." *Personality and Individual Differences*, 122 (2018): 29–37.

Jannini, E. A., A. Burri, P. Jern, and G. Novelli. 2015. "Genetics of Human Sexual Behavior: Where We Are, Where We Are Going." *Sexual Medicine Reviews* 3 (2): 65–77.

Oldmeadow, J. A., and B. J. Dixson. 2016. "The Association Between Men's Sexist Attitudes and Facial Hair." *Archives of Sexual Behavior* 45 (4): 891–899.

Reich, David. 2018. "How Genetics Is Changing Our Understanding of Race." https://www.nytimes.com/2018/03/23/opinion/sunday/genetics-race.html.

ECONOMIC BEHAVIORS

Camerer, C., G. Loewenstein, and D. Prelec. 2005. "Neuroeconomics: How Neuroscience Can Inform Economics." *Journal of Economic Literature* 43 (1): 9–64.

Glimcher, P. W., C. F. Camerer, E. Fehr and R. A. Poldrack. 2009. "Introduction: A Brief History of Neuroeconomics." In P. Glimcher and E. Fehr, eds., *Neuroeconomics*. New York: Academic Press, pp. 1–12.

Glimcher, P. W., and A. Rustichini. 2004. "Neuroeconomics: The Consilience of Brain and Decision." *Science* 306 (5695): 447–452.

INTELLIGENCE

Lee, J. J., R. Wedow, A. Okbay, E. Kong, O. Maghzian, M. Zacher, T. A. Nguyen-Viet, et al. 2018. "Gene Discovery and Polygenic Prediction from a Genome-wide Association Study of Educational Attainment in 1.1 Million Individuals." *Nature Genetics* 50 (8): 1112–1121.

Harden, K. P. 2018. "Why Progressives Should Embrace the Genetics of Education." https://www.nytimes.com/2018/07/24/opinion/dna-nature-genetics-education.html.

FIVE: HUMAN BEINGS AND CHOICE

Bernhard Rensch originally proposed Pyschozoa in 1959: the English translation was published 1972. Tattersall (2017) specifically addresses the issue of accelerated evolution in the human lineage. See deMenocal (2011) for a discussion of climatic influences on human evolution, and Gould (2002) and Coyne and Orr (2004) for the evolutionary effects of population disruption. Arranz-Otaegui, et al. (2018) reported the finding of Pleistocene bread. See Moore (2003) for details of Abu Hureyra, and other essays in Ammerman and Biagi (2003) for broader considerations of the Neolithic. Niles Eldredge quotation is from Eldredge (1995). For lactose tolerance see Harpending (2009). Green, et al. (2010) reported genomic introgression between Neanderthals and modern humans. For hominid brain enlargement trends see Tattersall (2012, 2018). See Gomulkiewicz and Kirkpatrick (1992) for a discussion of the reaction norm. Romain Gary's novel *Les Cerfs Volants*, quoted here and originally published in French in 1980, was published in English translation as *The Kites* in 2017. The Assyrian quote is cited from Patty and Johnson (1953), without guarantee. Other gloomy predictions are cited from Tattersall and Nevraumont (2018). Pinker references are to Pinker (2011), Piff reference is to Piff, et al. (2012), and Manne reference to Manne (2015). The *Guardian* quote can be found at: https://www.theguardian.com/commentisfree/2014/jul/08/the-age-of-entitlement-how-wealth-breeds-narcissism, and the *Economist* quote at: https://www.economist.com/middle-east-and-africa/2018/09/13/yemen-tries-gun-control.

Ammerman, A. J., and P. Biagi (eds). 2003. *The Widening Harvest. The Neolithic Transition in Europe: Looking Back, Looking Forward.* Boston: Archaeological Institute of America, pp. 59–74.

Arranz-Otaegui, A., L. G. Carretero, M. N. Ramsey, D. Q. Fuller, and T. Richter. 2018. "Archaeobotanical Evidence Reveals the Origins of Bread 14,400 Years Ago in Northeastern Jordan." *Proceedings of the National Academy of Sciences, USA* 115: 7925–7930.

Coyne, J. A., and H. A. Orr. 2002. *Speciation*. Sunderland, MA: Sinauer Associates.

DeMenocal, P. B. 2011. "Climate and Human Evolution." *Science* 331: 540–542.

Eldredge, N. 1995. *Dominion*. Berkeley: University of California Press.

Gary, Romain. 1980/2017. *The Kites*. New York: New Directions.

Gomulkiewicz, R., and M. Kirkpatrick. 1992. "Quantitative Genetics and the Evolution of Reaction Norms." *Evolution* 46 (2): 390–411.

Gould, S. J. 2002. *The Structure of Evolutionary Theory*. Cambridge, MA: Harvard University Press.

Green, R. E., J. Krause, J. A. W. Briggs, T. Maricic, et al. 2010. "A Draft Sequence of the Neandertal Genome." *Science* 328: 710–22.

Harpending, H. 2009. *The 10,000 Year Explosion: How Civilization Accelerated Human Evolution*. New York: Basic Books.

Manne, A. 2015. *The Life of I: The New Culture of Narcissism*. Melbourne: Melbourne University Publishing.

Moore, A. M. T. 2003. "The Abu Hureyra Project: Investigating the Beginning of Farming in Western Asia." In A. J. Ammerman and P. Biagi, eds., *The Widening Harvest. The Neolithic Transition in Europe: Looking Back, Looking Forward*. Boston, MA: Archaeological Institute of America, pp. 59–74.

Patty, W. L., and L. S. Johnson. 1953. *Personality and Adjustment*. New York: McGraw Hill.

Piff, P. K., D. M. Stancato, S. Cote, R. Mendoza-Denton, and D. Keltner. 2012. "Higher Social Class Predicts Increased Unethical Behavior." *Proceedings of the National Academy of Sciences, USA* 109: 4086–4091.

Pinker, S. 2011. *The Better Angels of Our Nature: Why Violence Has Declined*. New York: Penguin.

Rensch, B. 1972. *Homo sapiens: From Man to Demigod*. London: Methuen.

Tattersall, I. 2012. *Masters of the Planet: The Search for Our Human Origins*. New York: Palgrave Macmillan.

Tattersall, I. 2017. "Why Was Human Evolution So Rapid?" In A. Marom and E. Hovers, eds., *Human Paleontology and Prehistory: Essays in Honor of Yoel Rak*. Heidelberg: Springer, pp. 1–9.

Tattersall, I. 2018. "Brain Size and the Emergence of Modern Human Cognition." In J. H. Schwartz, ed., *Rethinking Human Evolution*. Cambridge, MA: MIT Press, pp. 319–334.

Tattersall, I., and P. Nevraumont. 2018. *Hoax. A History of Deception: 5,000 Years of Fakes, Forgeries, and Fallacies*. New York: Black Dog and Leventhal.

INDEX

O

P